Digital Communications 1

To all our families,
far and wide

Digital Communications 1

Fundamentals and Techniques

Safwan El Assad
Dominique Barba

iSTE

WILEY

First published 2020 in Great Britain and the United States by ISTE Ltd and John Wiley & Sons, Inc.

ISTE Ltd
27-37 St George's Road
London SW19 4EU
UK

www.iste.co.uk

John Wiley & Sons, Inc.
111 River Street
Hoboken, NJ 07030
USA

www.wiley.com

Library of Congress Control Number: 2020936357

British Library Cataloguing-in-Publication Data
A CIP record for this book is available from the British Library
ISBN 978-1-78630-541-1

Contents

Foreword

We have written this training book on digital communications in the spirit of presenting – in an integrated form – the basic knowledge on which modern digital communication systems are based and, above all, the way in which they are technically implemented, both in principle, and in practice. This book is the product of a long experience of training in this field in engineering school (Polytech Nantes, France).

The training is comprehensive: courses, tutorials presenting many standard problems targeted with detailed solutions, practical work concretely illustrating various aspects of the techniques of implementation.

As we have mentioned, although our experience is primarily that of training engineers, we have, through adaptations of the content, wished to address broader audiences: first in initial training, engineers, Master 2, specialized telecommunications licenses or other related specialties. But also to the trainers by providing them, through tutorials and practices (Lab Works), content that can be very useful in the construction of the training they provide. In continuing education, this book is also addressed to telecommunication technicians or for an additional year of specialization (specific years complementary to training in IUT).

This book, which is composed of two associated volumes, is presented in its first aspect, as a very concise and complete synthesis of the foundations and techniques of digital communications (Volume 1). It is broken down into two parts. The first part concerns the theory of information itself, which deals with both sources of information and communication channels, in terms of the errors they introduce in the transmission of information, as well as ways to protect the latter by using appropriate coding methods. The second part deals with the technical aspects of transmission, we first present the baseband transmission with the important concept of equalization and its implementations. The performance evaluation, in terms of probability of errors, is systematically developed and detailed as well as the on-line codes used. We then present the transmissions with digital modulation of carriers used in transmission (radio transmissions but also on electric cables).

A second important aspect, teaching knowledge and skills, composes this book (first part of Volume 2). It concerns the tutorial aspect of a course. This is an ordered set of about 30 standard problems with detailed solutions covering the different parts of the course. The set should allow a learner to gradually and deeply understand the essentials of this field and acquire the necessary skills to practice them in the industrial world.

Finally, the last aspect concerns practices in the proper sense of the term, an indispensable complement to training progressing to know-how (second part of Volume 2). We propose here a set of five lab works. The interest of these is that they go from the basic measurements on the transmission cables, to the design in software simulation of modems and cyclic coders, through the use of blocks of electronic modules carrying out basic functions useful in digital communications.

For every book sold, we will provide the buyer with two practical pieces of software from MATLAB-Simulink: “Modem QPSK” and “Cyclic encoder-decoder”, free of charge. We will provide necessary explanations and endeavor to help with the set-up of the two pieces of practical material.

Part 1

Theory of Information

Introduction to Part 1

Transmitting information between two entities, commonly referred to as the "source of information" for the first and the "recipient" or "user" for the second, assumes that this information is represented in a common and understandable form by both parties. It is also necessary that the amount of information exchanged does not exceed the transmission capacity of the channel. Lastly, it is desirable that the disturbances (modifications of all kinds) in the transmission channel have, if possible, little or, ideally, no effect on the information delivered to the recipient or user.

All this is studied, analyzed and treated in what is called the "theory of information", the theoretical basis of different techniques of communication.

In a schematic way, the problems raised and treated by this are the following:

– quantitative measurement of information;

– representation of common information between sender and recipient: code;

– source coding: adapting the source rate to the capacity of the transmission channel;

– channel coding: protecting against transmission errors.

The block diagram of Figure I1.1 shows at what levels of the communications chain these problems are, both on the side of the emission of the information (transmitter) as on the side of the reception of the information (receiver).

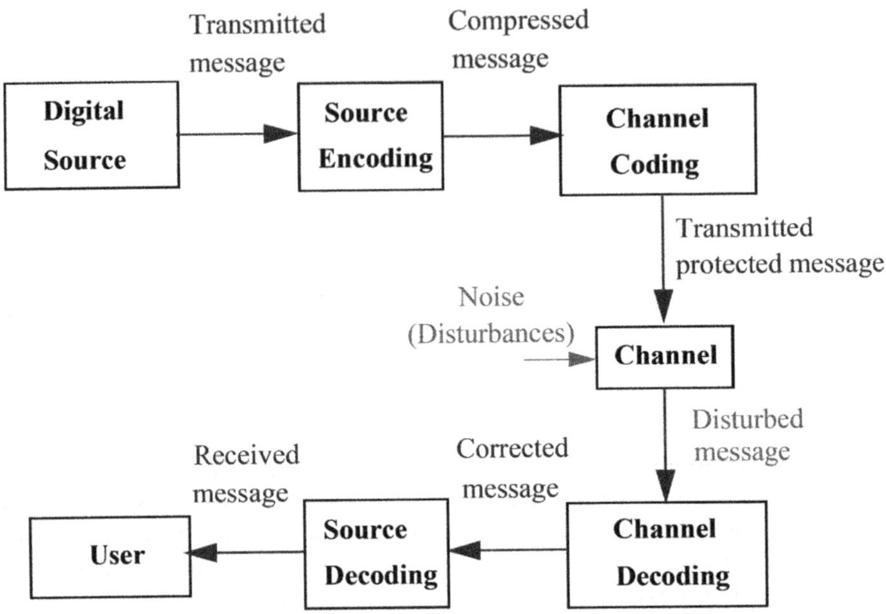

Figure I1.1. *General structure of a digital communications system and its problems. For a color version of this figure, see www.iste.co.uk/assad/digital1.zip*

We have divided this first part, dedicated to the theory of information, into three chapters following a first chapter of general introduction to telecommunications (Chapter 1). Each of these three chapters is devoted to each of the three issues raised previously:

– measurement of information (Chapter 2);

– source coding for the transmission on channels without disturbances (Chapter 3);

– channel coding for transmission on channels with disturbances (Chapter 4).

1

Introduction to Telecommunications

1.1. Role of a communication system

A communication system aims to convey as faithful and reliable messages as possible between a sender and a recipient, at any distance, with reasonable costs. Messages are information entities and their routing requires the existence of a communication channel to convey them (Figure 1.1).

Figure 1.1. *Simple communication system. For a color version of this figure, see www.iste.co.uk/assad/digital1.zip*

The particular features of a communication system are described below:

– the transportation of information (words, images, texts, data, etc.) and not the transportation of the information medium (paper, disk, etc.);

– the users of a telecommunications system want the messages they transmit to be delivered without loss or modification. This implies a high *fidelity* of the system (*quality*), in spite of the inevitable imperfections and disturbances to which it is subjected;

– the communication service must be available in all circumstances, irrespective of unforeseeable and unavoidable partial failures which can reach part of the system. This corresponds to a high *reliability* of the service (*availability*);

– the transportation of information at *various distances*, up to thousands of kilometers (or more) requires the use of specific transmission means (copper or optical cables, radio channels, satellites). This is a *transmission* problem.

1.1.1. *Types of services offered by communication systems*

We can characterize these in several ways by distinguishing (see Figure 1.2):

– the type of information transmitted;

– the number of partners;

– the role played by the partners: monologue, dialogue, conference.

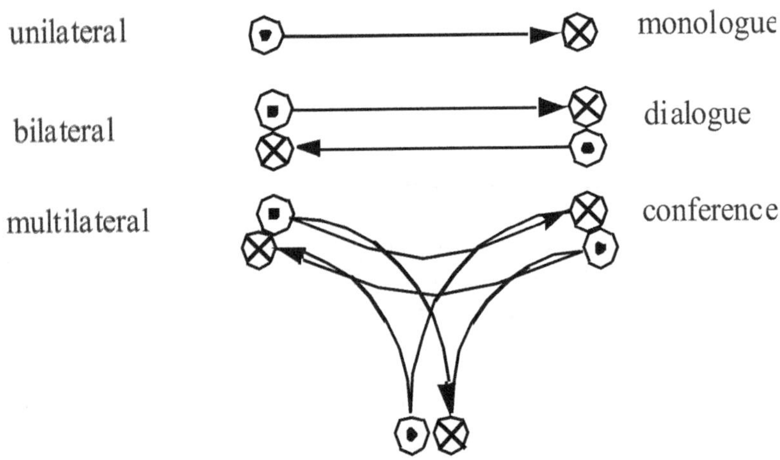

Figure 1.2. *Modes of communication: ⊙ source; ⊗ recipient*

When a certain number of users benefit from the same service, all the transmission links constitute a network. This corresponds to an information *broadcast network*, if one is dealing with the unilateral transmission of a source to several recipients, or to an information *collection network*, if the transmission is unilateral from several sources to only one recipient.

When there is no permanent transmission path between two points of a network, it is called a *switched transmission*: the link is established only after receipt and execution of connection orders from the partners. Generally, the switched network is very common, that is to say, the elements of the system relating to switching and transmission are common resources, made available to potential users who have access to this network by a means of individual transmission.

Figure 1.3 shows the three types of networks.

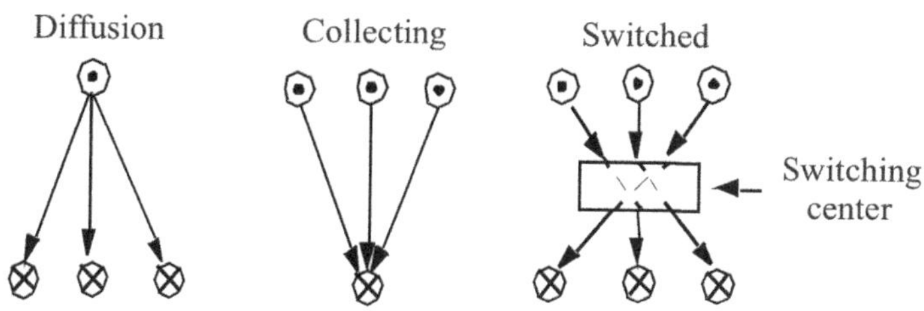

Figure 1.3. *Type of networks*

1.1.2. *Examples of telecommunications services*

We give some examples of telecommunications services in Table 1.1.

Information type	Communication mode	Network	Service
Speech	Bilateral Multilateral	Switched communication	Phone Conference call
Video	Unilateral	Fixed broadcast point to point	Video conference Television

Table 1.1. *Examples of telecommunications services*

The situation is not steady. In fact, the services offered and their diversity have considerably increased, both for long-distance telecommunications systems (in particular, for text, data, images and video-type information) and the considerable development of local business networks, which make it possible to connect computer and production equipment of a heterogeneous nature. Regarding telecommunications, almost all digital communications can integrate the transmission of different types of information in a single *ISDN* (Integrated Services Digital Network). In addition, communications over the Internet have been developing for almost a couple of decades.

1.2. Principle of communication

In a broad sense, the communication is a transfer of information (messages) between a sender and a recipient, through a medium called a channel. There are basically three main functions corresponding to this principle, as shown in Figure 1.4.

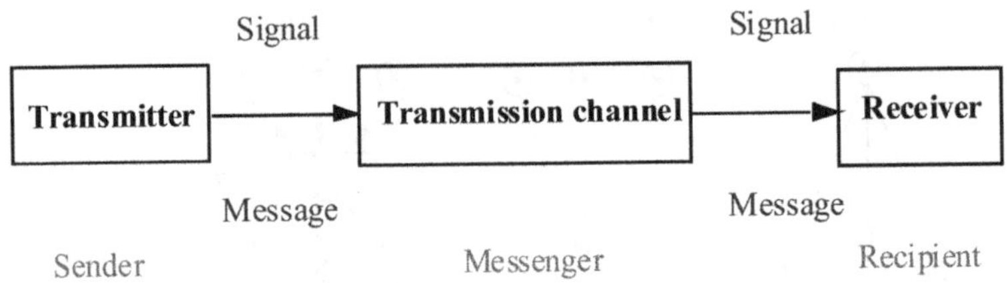

Figure 1.4. *Principle of communication. For a color version of this figure, see www.iste.co.uk/assad/digital1.zip*

A message has been considered as any means of modifying the state or evolution of the addressee. The signal is only an auxiliary supporting the message. It technically makes it possible to transmit this message.

The message has meaning only for the sender and the recipient. On the other hand, for the *messenger*, the meaning of the messages does not matter. Only the quantitative aspect interests them (to make money for the service means the transmission of a quantity of information). It is this approach that is used to design communication systems.

For communications, a message will be defined as a sequence of elements taken from a finite set of symbols: an alphabet. The transmission of a message consists of making a choice (according to a law that remains to be defined) among all the possible messages. The recipient is supposed to know, *a priori*, all possible messages. In addition, they must have a decision rule, allowing them to decide which message was sent. A more detailed diagram of a digital communication system makes it possible to highlight the various problems encountered in effectively achieving communication between the sender and the recipient of a message, because the message must satisfy the following conditions:

– the source and the recipient must agree on the symbolic representation used in the information transmitted (definition and use of a message representation code);

– the transmission channel must be transparent to the information, that is to say, it plays only a neutral conveyor role, without changing the information transmitted. This is only possible under certain constraints and by using special processing on the signal transmitted in the channel and on its reception;

– the information to be transmitted must be in a form compatible with the known characteristics of the available channel. This is done by prior coding.

This diagram corresponds to a digital information transmission system. The following elements are distinguished in Figure 1.5.

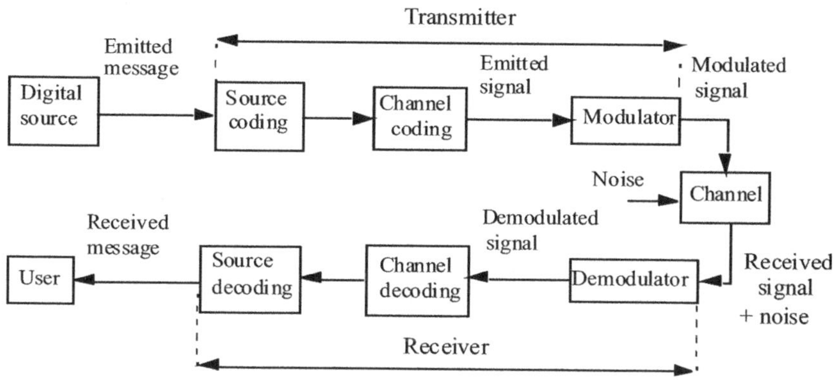

Figure 1.5. *Digital communication system*

The role of the different blocks is as follows:

– the source coding makes it possible to adapt a source whose bitrate is too high to the fixed capacity of the transmission channel (to minimize the redundancy of the source). It makes the best use of the channel capacity;

– channel coding helps, firstly, to protect against transmission errors (due to channel imperfections) by introducing a desired redundancy, and, secondly, to make a correspondence between a signal in the channel and one of the symbols of information. It is such that this signal must be compatible with the characteristics of the channel;

– the coding can also be used in the context of encryption (ciphering) of the information;

– modulators and demodulators are used when the message cannot be transmitted in baseband through the medium of transmission (due to propagation difficulties);

– the transmission channel itself transmits the transmitted signal to the receiver;

– the channel decoding restores the coded message (compressed, protected, etc.);

– source decoding reconstructs the original message;

– the information is finally processed by the recipient or stored.

In this system, the transmitted messages undergo *various types of degradations*:

– noise in the transmission channel disturbing the transmitted signal: noise, crosstalk, jamming, momentary fading, impulsive noise, etc., but also in the transmission and reception electronic devices;

– disturbances and distortions in the source coding for some information compression coding systems (audio, image, video);

– distortions created by the filtering made by the channel. These distortions translate into linear and non-linear distortions (non-constant group velocity). In addition, interferences between successive information symbols belonging to the same source of information or to different sources (in the case of time multiplexing) are developed and must be fought in a specific way.

All this leads to a very *important role played by the signal processing at different levels of the communication system*, especially at the receiver which regenerates the transmitted signal, before degradation in the transmission channel.

Optimizing the performance of communication systems has led to a diversification of methods and techniques used:

– a better adaptation of the signal emitted by an equalization pre-processing, associated with post-processing in the receiver, enables us to obtain a better robustness against noise of all kinds which is much greater than a single post-processing;

– the use of detector codes and/or error corrector codes permits us to transmit information in a very disturbed channel almost surely;

– the compression of information, by reducing the bitrate, makes it possible to increase the number of messages transmitted on the same transmission channel, sometimes at the cost of damage to the original message, but which can be controlled in such a way that the disturbances caused are insignificant (images, audio, etc.) and not perceived when they are used.

1.3. Trend towards digital communications

Previously, transmission systems used analog representation and transmission techniques almost exclusively. A fairly clear evolution has emerged since the 1970s for the introduction and use of digital techniques in representation and transmission, regardless of the nature of the information at the origin: digital (data) or analog (phone, images, and videos). In the latter case, this supposes the use of a process of digitization (sampling and quantization) of the signal.

This use of digital techniques in transmission is motivated by the following four reasons:

– degradations do not accumulate in digital transmissions (in the distortion sense);

– the possibility of regeneration upon reception of the transmitted signals with a probability of error such that the quality of transmission remains good;

– the cost of end equipment is lower mainly because of the integration of digital systems;

– monitoring the quality of the transmission and its control are easier.

This digitization was first applied to cable transmission systems and then to radio transmission systems: radio-relay systems and satellite telecommunications. The current trend is in the design and use of fully digital transmission systems (integrated transmission and switching) and the integration of services in the sense where in the same system, and therefore on the same transmission medium, a whole set of different services coexist.

However, a *digital transmission* a priori *requires a larger bandwidth than an analog transmission* if one use methods of compression and representation to no more than two levels (M-ary representation, with M> 2).

Indeed, an analog telephone channel conventionally occupies a bandwidth of 3.1 KHz, the same channel in digital form requires a bitrate of 64 kbit/s, so in binary transmission, a bandwidth of about 64 x 0.8 ≅ 51 KHz (according to the extended Nyquist criterion), that is, about 16 times more than analog transmission.

2

Measurement of Information of a Discrete Source and Channel Capacity

2.1. Introduction and definitions

The purpose of a message is to provide information to a recipient. Information, like energy, for that matter, is a fundamental notion, which is part of our daily life. It can be transported (transmitted), stored (memorized) and transformed (processed).

The value of information lies in the surprise effect it causes; it is all the more interesting because it is less predictable. As a result, information is assimilated to an event of a random nature, considered, in the following, as a random variable.

The measure of the information is then given by measuring the uncertainty of an event system, that is to say, the random choice of an event in a set of possible and distinct events.

We will use the terms *message* or *sequence* to indicate any finite sequence of symbols taken in an alphabet $A = \{\quad\}$ of a discrete source: a finite set of predetermined symbols, for example:

$$A_b = \{0, 1\}$$

$$A_{AES} = \{0, 1, 2, \cdots, 255\}$$

$$A_S = S = \{s_1, s_2, \cdots, s_N\}$$

We will use the term *discrete source of information* S_n to indicate the system selecting, from the set of symbols from $A_s = S$, the symbols to transmit:

$$A_s = S = \{s_1, s_2, \dots, s_N\}$$

Selection

$$S_n = [s_{i,1}, s_{j,2}, \dots, s_{m,n}] \quad i, j, m = 1, \dots, N$$

Sequence of symbols randomly issued — 1st symbol issued — nth symbol issued

The choice of $s_{m,n}$ is carried out according to a given probability law, which can be steady temporally or variable over time.

2.2. Examples of discrete sources

2.2.1. *Simple source (memoryless)*

This is a source for which the probability of the occurrence of a symbol does not depend on the previous symbols:

$$p(s_{k,n}/s_{i,1}, s_{j,2}, \cdots, s_{m,n-1}) = p(s_{k,n}) \quad i, j, k, m = 1, \cdots, N \qquad [2.1]$$

$$Pr\{S_n\} = p(s_{i,1}) \times p(s_{j,2}) \times \cdots \times p(s_{m,n}) \qquad [2.2]$$

2.2.2. *Discrete source with memory*

It is a source for which the probability of occurrence of a symbol depends on the symbols already emitted (statistical dependence) and on instants 1, 2, ..., n where they have been emitted:

$$Pr\{S_n\} = p(s_{i,1}) \times p(s_{j,2}/s_{i,1}) \times p(s_{k,3}/s_{i,1}, s_{j,2}) \times \cdots \times p(s_{m,n}/s_{i,1}, s_{j,2}, \cdots, s_{l,n-1}) \qquad [2.3]$$

If the source is in addition stationary then the dependence is only on the ranks and not on the moments when the symbols have been transmitted, so:

$$p(s_{i,n}) = p(s_{i,n+k}) \ \forall k \qquad [2.4]$$

that is, *the statistical properties are independent of any temporal translation.* We can consider that a text written in any language is an example of such a source.

2.2.3. *Ergodic source: stationary source with finite memory*

The ergodicity of an information source implies that a temporal average calculated over an infinite duration is equal to the statistical average.

2.2.4. *First order Markovian source (first order Markov chain)*

First order Markov sources play an important role in several domains, for example, in (Caragata *et al.* 2015), the authors used such a source to model the cryptanalysis of a digital watermarking algorithm.

It is characterized by:

$$p(s_{k,n}/s_{i,1}, s_{j,2}, \cdots, s_{l,n-1}) = p(s_{k,n}/s_{l,n-1}) \quad [2.5]$$

with:

$$i, j, k, l = 1, \cdots N; \quad s_{k,n} \in A_S = S = \{s_1, s_2, \dots, s_N\}$$

The probability $p(s_{k,n}/s_{l,n-1}) = p_{l,k}$ is called transition probability from state l to state k, and:

$$\sum_{k=1}^{N} p_{l,k} = 1 \quad [2.6]$$

The probability that at time n the source is in the state k is:

$$p(s_{k,n}) = \sum_{l=1}^{N} p(s_{k,n}, s_{l,n-1}) \quad \text{for } k = 1, \cdots, N$$

$$p(s_{k,n}) = \sum_{l=1}^{N} p(s_{l,n-1}) \times p(s_{k,n}/s_{l,n-1}) = \sum_{l=1}^{N} p(s_{l,n-1}) \times p_{l,k} \quad [2.7]$$

By introducing matrix notations:

$$\boldsymbol{P}_n = \begin{bmatrix} p(s_{1,n}) \\ p(s_{2,n}) \\ \cdots \\ p(s_{N,n}) \end{bmatrix}; \quad \boldsymbol{P}_{n-1} = \begin{bmatrix} p(s_{1,n-1}) \\ p(s_{2,n-1}) \\ \cdots \\ p(s_{N,n-1}) \end{bmatrix}; \quad \boldsymbol{T} = \begin{bmatrix} p_{11} & p_{12} & \cdots & p_{1N} \\ p_{21} & p_{22} & \cdots & p_{2N} \\ \cdots & \cdots & \cdots & \cdots \\ p_{N1} & p_{N2} & \cdots & p_{NN} \end{bmatrix} \quad [2.8]$$

Taking into account the relationship [2.8], the relation [2.7] is also written as:

$$\boldsymbol{P}_n = \boldsymbol{T}^t \times \boldsymbol{P}_{n-1} \quad [2.9]$$

where $\boldsymbol{T}^t$ is the transposed matrix of transition probabilities.

Moreover, if the source is, stationary, then:

$$\boldsymbol{P}_n = \boldsymbol{P}_{n-1}$$

in other words, *the probabilities of the occurrence of the symbols do not depend on n. It is the same for the transition probabilities* $p_{l,k}$, then the relation [2.9] is written:

$$\boldsymbol{P}_n = [\boldsymbol{T}^t]^n \times \boldsymbol{P}_0 \quad [2.10]$$

where $\boldsymbol{P}_0$ is the matrix of probabilities governing the generation of symbols by the source at the initial instant $n = 0$.

2.3. Uncertainty, amount of information and entropy (Shannon's 1948 theorem)

The realization of an event x of probability $p(x)$ conveys a quantity of information $h(x)$ related to its uncertainty. $h(x)$ is an increasing function of its improbability $1/p(x)$:

$$h(x) \cong f[1/p(x)]$$

If an event x is certain, then $p(x) = 1$, the uncertainty is zero and therefore the quantity of information $h(x)$ brought by its realization is null.

Moreover, let us consider the realization of a pair of independent events x and y. Their joint realization leads to the amount of information it brings being the sum of the quantities of information brought by each of them:

$$h(x, y) = h(x) + h(y) \cong f[1/p(x)] + f[1/p(y)] \quad [2.11]$$

but $h(x,y)$ is also written:

$$h(x,y) \cong f[1/p(x,y)] = f\left[\frac{1}{p(x)} \times \frac{1}{p(y)}\right] \quad [2.12]$$

so, from the relationships [2.11] and [2.12], the form of the function f is of logarithmic type, the base a of the logarithmic function can be any $a \in \Re^+$:

$$h(x) = \lambda \log_a[1/p(x)] = -\lambda \log_a[p(x)] \quad [2.13]$$

It is agreed by convention that we select, as the unit of measure of information, the information obtained by the random selection of a single event out of two equally probable events $p_i = p_j = 1/2$. In this case, we can write:

$$h(x_i) = h(x_j) = \lambda \log_a(2) = 1 \text{ } (\lambda\text{: a positive constant})$$

If we choose the logarithm in base 2, λ becomes equal to unity and therefore $h(x_i) = h(x_j) = \log_2(2) = 1$ Shannon (Sh) or 1 bit of information, not to be confused with the digital bit (*binary digit*) which represents one of the binary digits 0 or 1.

Finally, we can then write:

$$h(x) = -\log_2[p(x)] \quad [2.14]$$

It is sometimes convenient to work with logarithms in base e or with logarithms in base 10. In these cases, the units will be:

$\log_e e = 1$ natural unit = 1 nat (we choose 1 among e)

$\log_{10} 10 = 1$ decimal unit = dit (we choose 1 among 10)

Knowing that:

$$\log_a(x) = \log_e(x)/\log_e(a)$$

the relationships between the three units are:

– natural unit: 1 nat = $\log_2(e) = 1/\log_e(2) = 1.44$ bits of information;

– decimal unit: 1 dit= $\log_2(10) = 1/\log_{10}(2) = 3.32$ bits of information.

They are pseudo-units without dimension.

2.3.1. *Entropy of a source*

Let a stationary memoryless source S produce random independent events (symbols) s, belonging to a predetermined set $[S] = [s_1, s_2, \ldots, s_N]$. Each event (symbol) s_i is of given probability p_i, with:

$$\sum_{i=1}^{N} p_i = 1$$

The source S is then characterized by the set of probabilities $[P] = [p_1, p_2, \ldots, p_N]$. We are now interested in the average amount of information from this source of information, that is to say, resulting from the possible set of events (symbols) that it carries out, each is taken into account with its probability of occurrence. This average amount of information from the source S is called "*entropy* $H(S)$ *of the source*".

It is therefore defined by:

$$H(S) = E_S\{h(s_i)\} = \sum_{i=1}^{N} p_i h(s_i) = -\sum_{i=1}^{N} p_i \log_2 p_i \qquad [2.15]$$

2.3.2. *Fundamental lemma*

Let two probability partitions on S:

$$p_1, p_2, \cdots, p_N \text{ and } q_1, q_2, \cdots, q_N \text{ with } \sum_{i=1}^{N} p_i = \sum_{i=1}^{N} q_i = 1$$

we have the inequality:

$$\sum_{i=1}^{N} p_i \log_2 \left[\frac{q_i}{p_i}\right] \leq 0 \ (= 0 \text{ for } p_i = q_i) \qquad [2.16]$$

Indeed, since: $\log_e(x) \leq x - 1, \forall\, x$ positive real, then:

$$\sum_{i=1}^{N} p_i \log_e \left[\frac{q_i}{p_i}\right] \leq \sum_{i=1}^{N} p_i \left[\frac{q_i}{p_i} - 1\right] = \sum_{i=1}^{N} q_i - \sum_{i=1}^{N} p_i = 0$$

2.3.3. *Properties of entropy*

– *Positive*: since $0 \leq p_i \leq 1$; $\sum_{i=1}^{N} p_i = 1 \Rightarrow H > 0$ (with the agreement $p_i \log_2 p_i = 0$ for $p_i = 0$).

– *Continuous*: because it is a sum of continuous functions "log" of each p_i.

– *Symmetric*: relative to all the variables p_i.

– *Upper bounded: entropy has a maximum value*: $H(p_1, p_2, \cdots, p_N) \leq \log_2 N = H_{max}$ got for a uniform law: $p_i = 1/N$; $\forall\, i = 1, \ldots, N$

– *Additive*: let $p_1, \ldots, p_k$; $p_{k+1}, \ldots, p_N$ and $P = p_1 + p_2 + \cdots + p_k$; $Q = p_{k+1} + p_{k+2} + \cdots + p_N$; $(P + Q = 1)$, then:

$$H(p_1, \cdots, p_N) = H(P, Q) + PH\left(\frac{p_1}{P}, \cdots, \frac{p_k}{P}\right) + QH\left(\frac{p_{k+1}}{Q}, \cdots, \frac{p_N}{Q}\right) \quad [2.17]$$

2.3.4. *Examples of entropy*

2.3.4.1. *Two-event entropy (Bernoulli's law)*

$$H(S) = H(p_1, p_2) = -[p_1 \, log_2 \, p_1 + p_2 \, log_2 \, p_2]$$

$$p_2 = 1 - p_1 \rightarrow H(S) = H(p_1, 1 - p_1) = H(p_1)$$

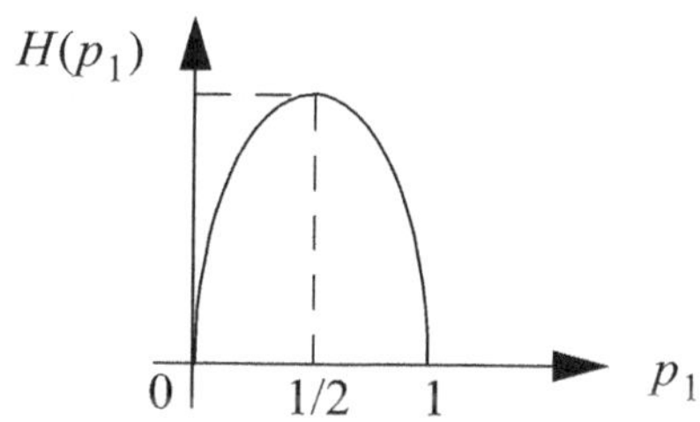

Figure 2.1. *Entropy of a two-event source*

The maximum of the entropy is obtained for $p_1 = 1/2$, equal to 1 bit of information.

2.3.4.2. *Entropy of an alphabetic source with (26 + 1) characters*

– *For a uniform law:*

$\Rightarrow H = \log_2(27) = 4.75$ bits of information per character

– In the French language (according to a statistical study):

$\Rightarrow H = 3.98$ bits of information per character

Thus, a text of 100 characters provides an information = 398 bits.

The inequality of the probabilities makes a loss of *475 – 398 = 77 bits of information.*

2.4. Information rate and redundancy of a source

The information rate of a source is defined by:

$$D_s = \frac{H(S)}{\bar{T}} \text{ (bit of information/s)} \qquad [2.18]$$

Where: $\bar{T}$ represents the average duration of a symbol emitted by the source.

The redundancy of a source is defined as follows:

$$\rho_s = H_{max}(S) - H(S) \qquad [2.19]$$

2.5. Discrete channels and entropies

Between the source of information and the destination, there is the medium through which information is transmitted. This medium, including the equipment necessary for transmission, is called the transmission channel (or simply the channel).

Let us consider a discrete stationary and memoryless channel (discrete: the alphabet of the symbols at the input and the one at the output are discrete).

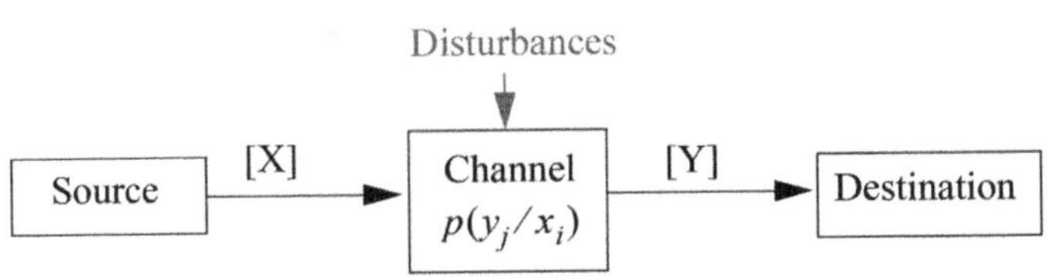

Figure 2.2. *Basic transmission system based on a discrete channel. For a color version of this figure, see www.iste.co.uk/assad/digital1.zip*

We denote:

– $[X] = [x_1, x_2, \ldots, x_n]$: the set of all the symbols at the input of the channel;

– $[Y] = [y_1, y_2, \ldots, y_m]$: the set of all the symbols at the output of the channel;

– $[P(X)] = [p(x_1), p(x_2), \ldots, p(x_n)]$: the vector of probability of symbols at the input of the channel;

– $[P(Y)] = [p(y_1), p(y_2), \ldots, p(y_m)]$: the vector of probability of symbols at the output of the channel.

Because of the perturbations, the space $[Y]$ can be different from the space $[X]$, and the probabilities $P(Y)$ can be different from the probabilities $P(X)$.

We define a product space $[X \cdot Y]$ and we introduce the matrix of the probabilities of the joint symbols, input-output $[P(X,Y)]$:

$$[P(X,Y)] = \begin{bmatrix} p(x_1,y_1) & p(x_1,y_2) & \cdots & p(x_1,y_m) \\ p(x_2,y_1) & p(x_2,y_2) & \cdots & p(x_2,y_m) \\ \cdots & \cdots & \cdots & \cdots \\ p(x_n,y_1) & p(x_n,y_2) & \cdots & p(x_n,y_m) \end{bmatrix} \quad [2.20]$$

We deduce, from this matrix of probabilities:

$$p(x_i) = \sum_{j=1}^{m} p(x_i, y_j) \quad [2.21]$$

$$p(y_j) = \sum_{i=1}^{n} p(x_i, y_j) \quad [2.22]$$

We then define the following entropies:

– the entropy of the source:

$$H(X) = -\sum_{i=1}^{n} p(x_i) \log_2 p(x_i) \quad [2.23]$$

– the entropy of variable Y at the output of the transmission channel:

$$H(Y) = -\sum_{j=1}^{m} p(y_j) \log_2 p(y_j) \quad [2.24]$$

– the entropy of the two joint variables (X, Y) input-output:

$$H(X,Y) = -\sum_{i=1}^{n}\sum_{j=1}^{m} p(x_i, y_j) \log_2 p(x_i, y_j) \quad [2.25]$$

2.5.1. *Conditional entropies*

Because of the disturbances in the transmission channel, if the symbol y_j appears at the output, there is an uncertainty on the symbol $x_i,\ i = 1, \ldots, . n$ which has been sent.

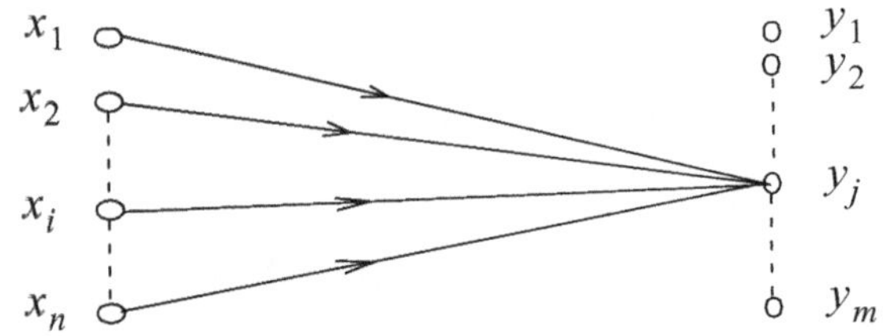

Figure 2.3. *Ambiguity on the symbol at the input when y_j is received*

The average value of this uncertainty, or the entropy associated with the receipt of the symbol y_j, is:

$$H(X/y_j) = -\sum_{i=1}^{n} p(x_i/y_j) \log_2 p(x_i/y_j) \qquad [2.26]$$

The mean value of this entropy for all the possible symbols y_j received is:

$$H(X/Y) = E_Y\{H(X/y_j)\} = \sum_{j=1}^{m} p(y_j)\, H(X/y_j) \qquad [2.27]$$

Which can be written as:

$$H(X/Y) = -\sum_{i=1}^{n}\sum_{j=1}^{m} p(y_j) \times p(x_i/y_j) \log_2 p(x_i/y_j) \qquad [2.28]$$

or:

$$H(X/Y) = -\sum_{i=1}^{n}\sum_{j=1}^{m} p(x_i, y_j) \log_2 p(x_i/y_j) \qquad [2.29]$$

The entropy $H(X/Y)$ is called equivocation (ambiguity) and corresponds to the *loss of information* due to disturbances (as $I(X,Y) = H(X) - H(X/Y)$). This will be specified a little further.

Because of disturbances, if the symbol x_i is issued, there is uncertainty about the received symbol y_j, $j = 1, \dots, m$.

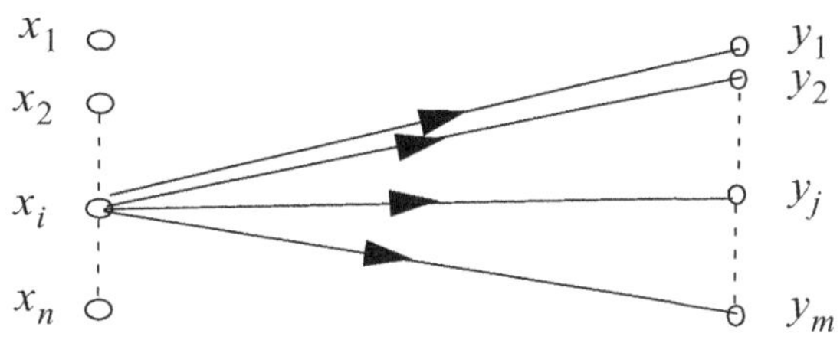

Figure 2.4. *Uncertainty on the output when we know the input*

The entropy of the random variable Y at the output knowing the X at the input is:

$$H(Y/X) = -\sum_{i=1}^{n}\sum_{j=1}^{m} p(x_i, y_j)\log_2 p(y_j/x_i) \quad [2.30]$$

This entropy is a measure of the uncertainty on the output variable when that of the input is known.

The matrix $P(Y/X)$ is called the *channel noise matrix*:

$$[P(Y/X)] = \begin{bmatrix} p(y_1/x_1) & p(y_2/x_1) & \cdots & p(y_m/x_1) \\ p(y_1/x_2) & p(y_2/x_2) & \cdots & p(y_m/x_2) \\ \dots & \dots & \dots & \dots \\ p(y_1/x_n) & p(y_2/x_n) & \cdots & p(y_m/x_n) \end{bmatrix} \quad [2.31]$$

A fundamental property of this matrix is:

$$\sum_{j=1}^{m} p(y_j/x_i) = 1 \text{ for each } i = 1, \cdots, n \quad [2.32]$$

Where: $p(y_j/x_i)$ is the probability of receiving the symbol y_j when the symbol x_i has been emitted.

In addition, one has:

$$p(x_i, y_j) = p(x_i) \times p(y_j/x_i) = p(y_j) \times p(x_i/y_j) \quad [2.33]$$

with:

$$p(y_j) = \sum_{i=1}^{n} p(x_i, y_j) = \sum_{i=1}^{n} p(x_i) \times p(y_j/x_i) \quad [2.34]$$

$p(y_j)$ is the probability of receiving the symbol y_j whatever the symbol x_i emitted, and:

$$p(x_i/y_j) = \frac{p(x_i) \times p(y_j/x_i)}{p(y_j)} \quad [2.35]$$

$p(x_i/y_j)$ is the probability that the symbol x_i was issued when the symbol y_j is received.

2.5.2. *Relations between the various entropies*

We can write:

$$H(X,Y) = -\sum_{i=1}^{n}\sum_{j=1}^{m} p(x_i, y_j) \log_2[p(x_i) \times p(y_j/x_i)]$$

$$H(X,Y) = -\sum_{i=1}^{n}\left[\sum_{j=1}^{m} p(x_i, y_j)\right] \log_2 p(x_i) - \sum_{i=1}^{n}\sum_{j=1}^{m} p(x_i, y_j) \log_2 p(y_j/x_i)$$

$$H(X,Y) = H(X) + H(Y/X) \quad [2.36]$$

In the same way, as one has: $H(Y,X) = H(X,Y)$, therefore:

$$H(X,Y) = H(Y) + H(X/Y) \quad [2.37]$$

In addition, one has the following inequalities:

$$0 \le H(X/Y) \le H(X) \le H(X,Y) \le H(X) + H(Y) \le 2H(X,Y) \quad [2.38]$$

and similarly:

$$0 \le H(Y/X) \le H(Y) \le H(X,Y) \le H(X) + H(Y) \le 2H(X,Y) \quad [2.39]$$

SPECIAL CASES.–

– *Noiseless channel:* in this case, on receipt of y_j, there is certainty about the symbol actually transmitted, called x_i (one-to-one correspondence), therefore:

$$p(x_i/y_j) = 1 = p(y_j/x_i) \quad [2.40]$$

Consequently:

$$H(X/Y) = H(Y/X) = 0 \quad [2.41]$$

and:

$$H(X,Y) = H(X) = H(Y) \quad [2.42]$$

– *Channel with maximum power noise:* in this case, the variable at the input is independent of that of the output, i.e.:

$$p(x_i/y_j) = p(x_i) \ ; \ p(y_j/x_i) = p(y_j) \ ; \ p(x_i, y_j) = p(x_i) \times p(y_j) \quad [2.43]$$

We then have:

$$H(X/Y) = H(X) \quad [2.44]$$

$$H(Y/X) = H(Y) \quad [2.45]$$

$$H(X,Y) = H(X) + H(Y) \quad [2.46]$$

NOTE.– *In information security,* if x_i is the plaintext, and y_j is the corresponding ciphertext, then $p(x_i/y_j) = p(x_i)$ is the condition of the perfect secret of a cryptosystem.

2.6. Mutual information

The mutual information obtained on the symbol x_i when the symbol y_j is received is given by:

$$\begin{aligned} i(x_i, y_j) &= -\log_2 p(x_i) - [-\log_2 p(x_i/y_j)] = \log_2 \left[\frac{p(x_i/y_j)}{p(x_i)}\right] \\ &= \log_2 \left[\frac{p(x_i, y_j)}{p(x_i) \times p(y_j)}\right] \end{aligned} \quad [2.47]$$

where $i(x_i, y_j)$ represents the decrease in uncertainty on x_i due to the receipt of y_j.

The average value of the mutual information, or the amount of information $I(X,Y)$ transmitted through the channel is:

$$I(X,Y) = \sum_{i=1}^{n}\sum_{j=1}^{m} p(x_i, y_j) \log_2 \left[\frac{p(x_i, y_j)}{p(x_i) \times p(y_j)}\right] \quad [2.48]$$

or:

$$I(X,Y) = -\sum_{i=1}^{n}\left[\sum_{j=1}^{m} p(x_i, y_j)\right] \log_2 p(x_i) - \sum_{j=1}^{m}\left[\sum_{i=1}^{n} p(x_i, y_j)\right] \log_2 p(y_j)$$

$$+\sum_{i=1}^{n}\sum_{j=1}^{m} p(x_i, y_j) \log_2 p(x_i, y_j) \quad [2.49]$$

Hence:

$$I(X,Y) = H(X) + H(Y) - H(X,Y) \quad [2.50]$$

Given relationships [2.36] and [2.37], we get:

$$I(X,Y) = H(X) - H(X/Y) \quad [2.51]$$

$$I(X,Y) = H(Y) - H(Y/X) \quad [2.52]$$

$$I(X,Y) = I(Y,X) \geq 0 \quad [2.53]$$

SPECIAL CASES.–

– *Noiseless channel*: X and Y symbols are linked, so:

$$I(X,Y) = H(X) = H(Y)$$

– *Channel with maximum power noise*: X and Y symbols are independent, therefore:

$$I(X,Y) = 0$$

2.7. Capacity, redundancy and efficiency of a discrete channel

Claude Shannon introduced the concept of channel capacity, to measure the efficiency with which information is transmitted, and to find its upper limit.

The *capacity C of a channel: (information bit/symbol)* is the maximum value of the mutual information $I(X,Y)$ over the set of input symbols probabilities $\{p(x_i)\}_{i=1,\cdots,n}$.

$$C = \underset{\{p(x_i)\}_{i=1,\cdots,n}}{Max}\, I(X,Y) = \underset{\{p(x_i)\}_{i=1,\cdots,n}}{Max}\, [H(Y) - H(Y/X)] \qquad [2.54]$$

The maximization of $I(X,Y)$ is performed under the constraints that:

$$p(x_i) \geq 0\,; \sum_{i=1}^{n} p(x_i) = 1$$

The maximum value of $I(X,Y)$ occurs for some well-defined values of these probabilities, which thus define a certain so-called *secondary source*.

The *capacity of the channel* can also be related *to the unit of time (bitrate C_t of the channel)*, in this case, one has:

$$C_t = \frac{C}{T} \text{ (bit of information/s)} \qquad [2.55]$$

The *channel redundancy* R_c and the *relative channel redundancy* ρ_c are defined by:

$$R_c = C - I(X,Y) \qquad [2.56]$$

$$\rho_c = 1 - \frac{I(X,Y)}{C} \qquad [2.57]$$

The *efficiency of the use of the channel* η_c is defined by

$$\eta_c = \frac{I(X,Y)}{C} = 1 - \rho_c \qquad [2.58]$$

2.7.1. *Shannon's theorem: capacity of a communication system*

Shannon also formulated the capacity of a communication system by the following relation:

$$C_t = B \times \log_2\left[1 + \frac{P_s}{B\Gamma_{b_0}}\right] \quad [2.59]$$

where:

- B: is the channel bandwidth, in hertz;

- P_s: is the signal power, in watts;

- Γ_{b_0}: is the power spectral density of the (supposed) Gaussian and white noise in its frequency band B;

- $B\Gamma_{b_0}$: is the noise power, in watts.

EXAMPLE.– Binary symmetric channel (BSC).

Any binary channel will be characterized by the noise matrix:

$$[P] = [P(Y/X)] = \begin{bmatrix} p(y_1/x_1) & p(y_2/x_1) \\ p(y_1/x_2) & p(y_2/x_2) \end{bmatrix}$$

If the binary channel is symmetric, then one has:

$$p(y_1/x_2) = p(y_2/x_1) = p$$

$$p(y_1/x_1) = p(y_2/x_2) = 1 - p$$

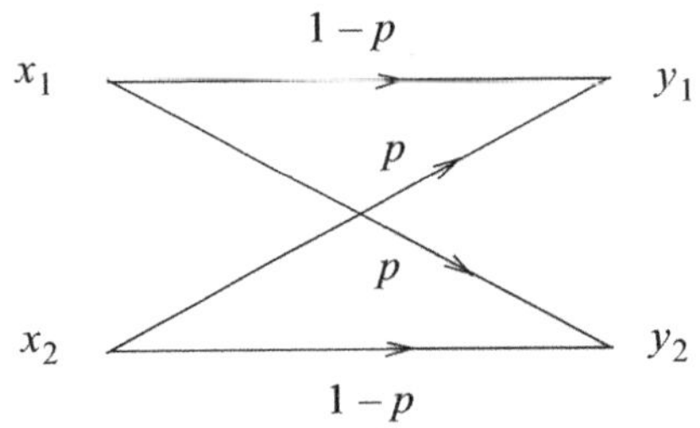

$$A_b \Rightarrow \begin{Bmatrix} x_1 = y_1 = 0 \\ x_2 = y_2 = 1 \end{Bmatrix}$$

Figure 2.5. *Binary symmetric channel*

The channel capacity is:

$$C = \underset{\{p(x_i)\}_{i=1,\cdots,n}}{Max}\,[H(Y) - H(Y/X)]$$

The conditional entropy $H(Y/X)$ is:

$$H(Y/X) = -\sum_{i=1}^{2}\sum_{j=1}^{2} p(x_i) \times p(y_j/x_i) \times \log_2 p(y_j/x_i)$$

or:

$$H(Y/X) = -\{[p(x_1) + p(x_2)][p\log_2 p + (1-p)\log_2(1-p)]\}$$

$$H(Y/X) = -[p\log_2 p + (1-p)\log_2(1-p)] = H(p)$$

Hence:

$$C = \underset{\{p(x_i)\}_{i=1,\cdots,n}}{Max} H(Y) - H(p)$$

But $\max H(Y) = 1$ for $p(y_1) = p(y_2)$. It follows from the symmetry of the channel that if $p(y_1) = p(y_2)$, then $p(x_1) = p(x_2) = 1/2$, and C will be given by:

$$C = 1 + p\log_2 p + (1-p)\log_2(1-p) = 1 - H(p)$$

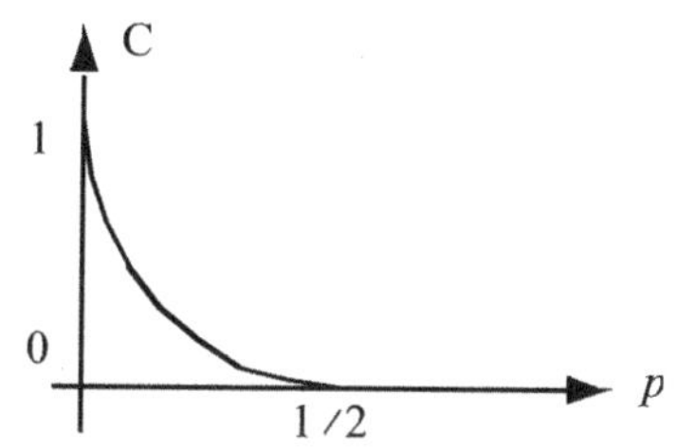

Figure 2.6. *Variation of the capacity of a BSC according to p*

2.8. Entropies with k random variables

The joined entropy of k random variables is written:

$$H(X_1, X_2, \cdots, X_k) = H(X_1) + H(X_2/X_1) + H(X_3/X_1, X_2) + \ldots$$

$$+ H(X_k/X_1, X_2, X_3, \cdots, X_{k-1})$$

$$H(X_1, X_2, \cdots, X_k) = \sum_{i=1}^{k} H(X_i/X_1, X_2, X_3, \cdots, X_{i-1}) \quad [2.60]$$

Furthermore, by setting down in the relationship:

$$H(X/Y) \leq H(X)\ ;\ X = X_i \text{ and } Y = X_1, X_2, \cdots, X_{i-1}$$

One has:

$$\sum_{i=1}^{k} H(X_i/X_1, X_2, \cdots, X_{i-1}) \leq \sum_{i=1}^{k} H(X_i) \qquad [2.61]$$

Equality occurs when the variables are independent.

3

Source Coding for Non-disturbance Channels

3.1. Introduction

The purpose of source coding for non-disturbance channels is:

– to transcribe the symbols of the source in the form of a sequence of letters taken from a predefined alphabet, making it possible to easily elaborate the signal that will carry the information, or to easily handle this information automatically. For this, different codes for information representation are used;

– to minimize redundancy of the source;

– to generate maximum mutual information: a statistical adaptation of the source to the channel.

We will now study the performance of some important codes.

3.2. Interest of binary codes

To show the interest of binary codes, we will take the example of coding a message in French. We can consider the following solutions:

– number every word in the dictionary, i.e. in number N and then assign *one amplitude out of N* of a signal to each word;

– assign *one amplitude out of 27 of a signal* to each letter of the alphabet (26 + 1);

– use five binary digits to encode numbers from 0 to 31, each of which represents a letter of the alphabet. In this case, it is sufficient to use a *two-level signal*, which is easily achievable, but at the cost of a longer decoding time in reception.

3.3. Single decoding codes

Let us take a discrete memoryless source which provides symbols taken on a set $[S]$ of symbols s called the source alphabet:

$$[S] = [s_1, s_2, \dots, s_N]$$

Whose probabilities of use by the source are:

$$[P] = [p(s_1), p(s_2), \dots, p(s_N)]$$

Let $[X]$ be the finite alphabet of the code (hence of the channel):

$$[X] = [x_1, x_2, \dots, x_q] \ (q = 2 \text{ for a binary code})$$

With these letters of the alphabet, we form N codewords (code C):

$$[C] = [c_1, c_2, \dots, c_N]$$

Each codeword c_k consists of a finite sequence of letters from the alphabet $[X]$:

$$c_k = x_i, x_j, \dots, x_k \qquad i, j, k \in [1, 2, \dots, q] \tag{3.1}$$

For $q = 2$, the alphabet is binary.

Coding is the operation of performing a one-to-one correspondence between the symbols (or sequence of symbols) $s_k \in [S]$ and the codewords $c \in [C]$.

EXAMPLE.– Let us take a text formed of n symbols taken on $S_n = [s_{i,1}, s_{j,2}, \dots, s_{l,n}]$ with $i, j, l = 1, \dots, N$, its coding is $C_n = [c_{i,1}, c_{j,2}, \dots, c_{l,n}]$.

Decoding implies that one is able *to separate the codewords in the received sequence uniquely*. This leads to the use of *regular codes and unique decoding* (known as "*decodable codes*").

When a single decoding code does not use any delimiter between words, it is called "separable code".

In the following, we will give the classification of the codes and some examples of single and separable decoding codes (with the exception of codes C1 and C2), some properties of which will be verified.

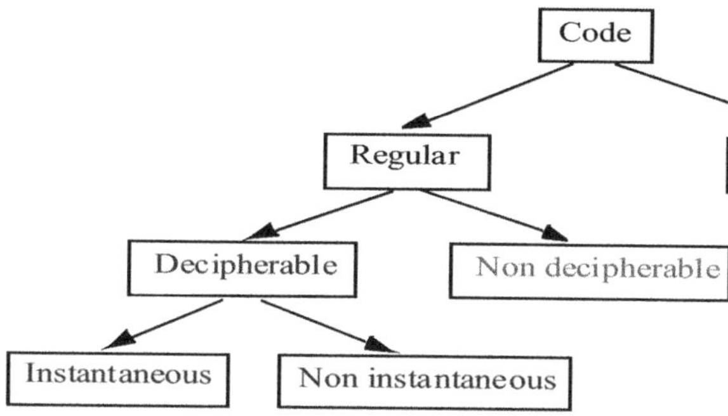

Figure 3.1. *Classification of codes. For a color version of this figure, see www.iste.co.uk/assad/digital1.zip*

Symbol s_k	Code C1	Code C2	Code C3	Code C4	Code C5	Code C6
s_1	1	0	00	0	0	0
s_2	1	10	01	10	01	10
s_3	0	00	10	110	011	110
s_4	00	01	11	1110	0111	111

Table 3.1. *Examples of unique decoding and separable codes (except codes C1 and C2)*

3.3.1. *Regular code*

A code is regular if each of its codewords is associated with only one of symbols $s_i \neq s_j \Longrightarrow c_i \neq c_j$. Code C1 is not a regular code.

3.3.2. *Single-decoded code (decipherable or decodable code)*

A code is single-decoded if all different texts lead to separate encodings:

$$\text{if } \left[s_{i,1}, s_{j,2}, \dots, s_{l,n}\right] \neq \left[\acute{s}_{i,1}, \acute{s}_{j,2}, \dots, \acute{s}_{l,k}\right] \quad \forall\, i, j, k, l$$

$$\text{then } \left[c_{i,1}, c_{j,2}, \dots, c_{l,n}\right] \neq \left[\acute{c}_{i,1}, \acute{c}_{j,2}, \dots, \acute{c}_{l,k}\right]$$

Code C2 is not a decipherable code. Indeed, let 010100 be the coded sequence. This sequence, read from the left, can represent four different sequences of symbols:

$$0\,|\,1\,0\,|\,1\,0\,|\,0 \quad \rightarrow \quad s_1\; s_2\; s_2\; s_1$$

$$0\,1\,|\,0\,1\,|\,0\,0 \quad \rightarrow \quad s_4\; s_4\; s_3$$

$$0\,1\,|\,0\,1\,|\,0\,|\,0 \quad \rightarrow \quad s_4\; s_4\; s_1\; s_1$$

$$0\,1\,|\,0\,|\,1\,0\,|\,0 \quad \rightarrow \quad s_4\; s_1\; s_2\; s_1$$

Codes C3, C4, C5 and C6 are decodable for the following reasons:

– C3 is a fixed-length code;

– C4 is a code with a 0 symbol indicating the end of each codeword. In addition, it is instantaneous;

– C5 is a code with a 0 symbol which indicates the beginning of each codeword. It is not instantaneous;

– C6 is an instantaneous code, so it is decipherable. The converse is not always true.

3.3.3. *Instantaneous code (irreducible code)*

This is a decipherable code that can be read directly without any particular artifice (fixed-length code, code using a separator that lengthens the text). For this, the code must check the condition of the prefix.

3.3.4. *Prefix*

Let $c_m = x_i, x_j, \ldots, x_m$ be a codeword, with: $i, j, m \in [1, 2, \ldots, q]$.

The *sequence of letters* $[x_i, x_j, \ldots, x_k]$ *where* $k < m$ *is called the prefix of the codeword* c_m.

PREFIX PROPERTY.– The necessary and sufficient condition for a code to be instantaneous is that no codeword is prefixed to another codeword.

EXAMPLE.– Let us take the following sequence, coded: 0 1 0 1 1 0.

– With code C4, as one receives the succession of letters of the alphabet of the code, one can determine the words of the code without referring to the following letters. Thus:

$$0 \mid 1\,0 \mid 1\,1\,0 \mid \rightarrow s_1 \;\; s_2 \;\; s_3$$

– With code C5, when decoding it is necessary that, after the reception of a word of the code, one also receives some of the following letters, to know which word was transmitted. So:

0 1 0 1 1 0

We cannot take any decision on receipt of the first 0 because, besides symbol s_1, we could also have the symbols s_2 or s_3 or s_4 and, for this reason, the decoding will be delayed until the receipt of the following letters, namely 1 and 0. It is only after the appearance of the 0 that one is sure that the only possible interpretation is 01, namely s_2 and so on:

$$0\,1 \mid 0\,1\,1 \mid 0 \rightarrow s_2 \;\; s_3 \;\; s_1$$

3.3.5. *Design of an instantaneous binary code*

We can design an instantaneous code from the following algorithm:

– divide the set $[S]$ of symbols of the source into two subsets S_0 and S_1:

$$S_0 = s_1, s_2, \ldots, s_k \;\; ; S_1 = s_{k+1}, s_{k+2}, \ldots, s_N$$

– assign the bit "0" to all the symbols of S_0 and the bit "1" to all the symbols of S_1;

– repeat the same operation on S_0 and S_1 and so on;

– stop the process when each subset has only one symbol s_i.

The code thus formed is instantaneous because no codeword is the prefix of another codeword.

EXAMPLE.–

<table>
<tr><td>s_1</td><td rowspan="4">0</td><td rowspan="2">0</td><td>0</td><td>000</td></tr>
<tr><td>s_2</td><td>1</td><td>001</td></tr>
<tr><td>s_3</td><td rowspan="2">1</td><td>0</td><td>010</td></tr>
<tr><td>s_4</td><td>1</td><td>011</td></tr>
<tr><td>s_5</td><td rowspan="4">1</td><td rowspan="2">0</td><td>0</td><td>100</td></tr>
<tr><td>s_6</td><td>1</td><td>101</td></tr>
<tr><td>s_7</td><td rowspan="2">1</td><td>0</td><td>110</td></tr>
<tr><td>s_8</td><td>1</td><td>111</td></tr>
</table>

Table 3.2. *Example of the design of an instantaneous binary code*

3.3.6. *Kraft–McMillan inequality*

It is often interesting to have a test to detect the existence of an instantaneous code. This test is established by the following theorem.

A necessary and sufficient condition for the existence of an instantaneous code of N words of lengths $[L] = [l_1, l_2, \ldots, l_N]$ coded on the finite X alphabet of q letters is:

$$\sum_{i=1}^{N} q^{-l_i} \leq 1 \qquad [3.2]$$

3.4. Average codeword length

3.4.1. *Coding efficiency in terms of transmission speed*

Let:

– τ be the transmission time of a letter x_i of the alphabet $[X]$ of the code $\forall\, i = 1, \ldots, q$;

– $t_i = l_i \tau$ be the transmission time of a codeword c_i.

Without losing generality, one can take $\tau = 1$, then: $t_i = l_i$.

The average duration $\bar{t}$ of transmission of a codeword is then:

$$\bar{t} = \sum_{i=1}^{N} p(c_i)t_i = \sum_{i=1}^{N} p(s_i)l_i = \bar{l} \quad [3.3]$$

The shorter the length $\bar{l}$, the better the transmission time $\bar{t}$.

3.4.2. *Minimum average codeword length $\overline{l_{min}}$*

The entropy of the source is:

$$H(S) = -\sum_{i=1}^{N} p(s_i)\log_2 p(s_i) \quad [3.4]$$

For an optimal coding of the source, the entropy $H(C)$ of the codewords is equal to $H(S)$.

The entropy of the code alphabet $[X]$ is:

$$H(X) = -\sum_{i=1}^{q} p(x_i)\log_2 p(x_i) \quad [3.5]$$

The average amount of information per codeword is given by the product of the average number of letters $\bar{l}$ and the average amount of information per letter (X):

$$H(C) = \bar{l} \times H(X) \quad [3.6]$$

Furthermore:

$$H(X) \leq \log_2 q = H_{max}(X) \quad [3.7]$$

(equality when $p(x_i) = p(x_j)\ \forall\ i,j\ \in [1,\cdots,q], i \neq j$).

Given this inequality, the relationship [3.6] becomes:

$$H(C) = \bar{l}H(X) \leq \bar{l}\log_2 q = \bar{l}H_{max}(X) \quad [3.8]$$

and since $H(S) = H(C)$, we have:

$$\bar{l} \geq \frac{H(S)}{\log_2 q} = \overline{l_{min}} \qquad [3.9]$$

In the case of a binary alphabet ($q = 2$), the relation [3.9] becomes:

$$\bar{l} \geq H(S) = \overline{l_{min}} \qquad [3.10]$$

So here, the entropy of the source represents the minimum average length $\overline{l_{min}}$ of a binary codeword.

3.5. Capacity, efficiency and redundancy of a code

The capacity of the code is the maximum value of the entropy of the code alphabet:

$$C = \max H(X) = \log_2 q \qquad [3.11]$$

The efficiency of a code is defined by:

$$\eta = \frac{\overline{l_{min}}}{\bar{l}} = \frac{H(S)}{\bar{l}\log_2 q} = \frac{H(X)}{\log_2 q} \qquad [3.12]$$

The redundancy of a code is then:

$$\rho = 1 - \eta = \frac{\log_2 q - H(X)}{\log_2 q} \qquad [3.13]$$

3.6. Absolute optimal codes

Absolute *optimal codes* are characterized by efficiency equal to unity. This means that the average length $\bar{l}$ takes as its minimum value $\overline{l_{min}}$. This is achieved when the letters of the code alphabet are used with equal probabilities:

$$p(x_1) = p(x_2) = \cdots = p(x_q) = 1/q \qquad [3.14]$$

The letters of the code alphabet are considered independent, hence:

$$p(s_i) = p(c_i) = [1/q]^{l_i} = q^{-l_i} \qquad [3.15]$$

Or, since: $\sum_{i=1}^{N} p(s_i) = 1$, it follows that:

$$\sum_{i=1}^{N} q^{-l_i} = 1 \quad [3.16]$$

If a code verifies the relation [3.16], then it is an absolute optimal code.

3.7. K-order extension of a source

Let us take a source $[S] = [s_1, s_2, \dots, s_N]$. We call the extension of order k of this source the source $[S^k]$, of which each symbol is obtained by the juxtaposition of k symbols of the source $[S]$.

We consider that the random source $[S]$ generates successive independent symbols. In this case, the probability of each symbol of the source $[S^k]$ is then equal to the product of the probabilities of the k symbols of the source $[S]$, and the entropy of the source $[S^k]$ is: $H(S^k) = k \times H(S)$ (see section 3.7.1 for the demonstration).

3.7.1. *Entropy of the 2nd order extension of a source* $[S]$

Let us take a source $[S] = [s_1, s_2, \dots, s_N]$, with the set $[P]$ of probabilities:

$[P] = [p(s_1), p(s_2), \dots, p(s_N)]$.

The entropy of the source $[S]$ is:

$$H(S) = -\sum_{i=1}^{N} p(s_i) \log_2 p(s_i)$$

Let $[S^2]$ be the second order extension of the source $[S]$. Symbols s_{ij} of the extension belong to a two-dimensional space, such as:

$$[S^2] = \begin{bmatrix} s_{11} & s_{12} & \cdots & s_{1N} \\ s_{21} & s_{22} & \cdots & s_{2N} \\ \cdots & \cdots & \cdots & \cdots \\ s_{N1} & s_{N2} & \cdots & s_{NN} \end{bmatrix} = \begin{bmatrix} s_1 s_1 & s_1 s_2 & \cdots & s_1 s_N \\ s_2 s_1 & s_2 s_2 & \cdots & s_2 s_N \\ \cdots & \cdots & \cdots & \cdots \\ s_N s_1 & s_N s_2 & \cdots & s_N s_N \end{bmatrix} \quad [3.17]$$

and:

$$[P(S^2)] = \begin{bmatrix} p(s_{11}) & p(s_{12}) & \cdots & p(s_{1N}) \\ p(s_{21}) & p(s_{22}) & \cdots & p(s_{2N}) \\ \cdots & \cdots & \cdots & \cdots \\ p(s_{N1}) & p(s_{N2}) & \cdots & p(s_{NN}) \end{bmatrix} \quad [3.18]$$

With, for a source producing independent symbols:

$$p(s_{ij}) = p(s_i) \times p(s_j) \quad [3.19]$$

The entropy $H(S^2)$ of the source $[S^2]$ can be written:

$$H(S^2) = -\sum_{i=1}^{N}\sum_{j=1}^{N} p(s_{ij}) \log_2 p(s_{ij}) \quad [3.20]$$

Let us develop the expression:

$$H(S^2) = -\sum_{i=1}^{N}\sum_{j=1}^{N} p(s_i) \times p(s_j)[\log_2 p(s_i) + \log_2 p(s_j)]$$

$$H(S^2) = -\sum_{j=1}^{N} p(s_j) \sum_{i=1}^{N} p(s_i) \log_2 p(s_i) - \sum_{i=1}^{N} p(s_i) \sum_{j=1}^{N} p(s_j) \log_2 p(s_j) \quad [3.21]$$

$$= 2H(S)$$

For a k-order extension, the expression [3.21] becomes:

$$H(S^k) = k \times H(S) \quad [3.22]$$

3.7.2. *Simple example of the interest of coding a source extension*

Let us take a two-symbol source, presented in Table 3.3.

Symbol	Probability	Code	$H(S) = 0.811$ bits of information/symbol
s_1	1/4	0	$\bar{l} = 1$ bit/coded symbol
s_2	3/4	1	$\eta = 81.1\,\%$ and $\rho = 18.9\,\%$

Table 3.3. *Characteristics of a two-symbol source*

The 2nd order extension of the source and the Huffman coding of this source (see section 3.9.2) are:

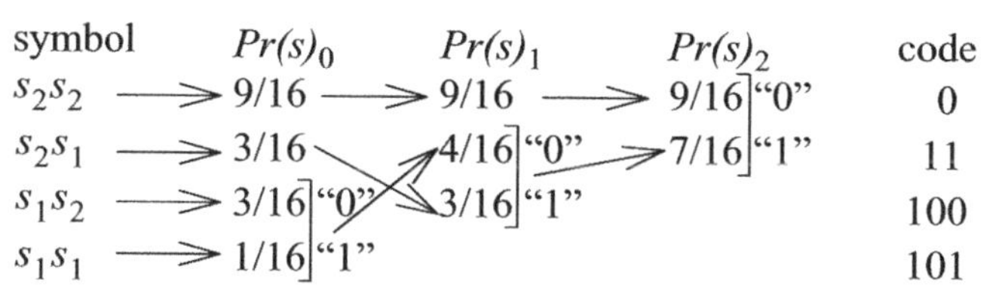

$$H(S^2) = 2H(S) = 1.622 \text{ bit of information/codeword}$$

$$\bar{l_2} = \sum_{i=1}^{4} p(c_i)l_i = 1.69 \text{ bit/codeword}$$

$$\eta_2 = H(S^2)/\bar{l_2} \cong 96\,\%\,;\ \rho \cong 4\,\%$$

$$\bar{l} = \bar{l_2}/2 = 0.845 \text{ bit/coded symbol}$$

These results show the interest of k-order extension of a source (here, with: $k = 2$).

3.8. Shannon's first theorem

STATEMENT.– *For any stationary source, there exists a uniquely decodable code, such that the average length $\bar{n}$ per symbol is as close as possible to its lower bound $H(S)/\log q$.*

Indeed, since the messages to be coded have an arbitrary set of probabilities (not satisfying [3.15]), we can easily show, from the logarithm of [3.15] and some developments that:

$$\frac{H(S)}{\log_2 q} \le \bar{l} < \frac{H(S)}{\log_2 q} + 1 \qquad [3.23]$$

This relationship is valid for any memoryless source. It is particularly valid for any extension $[S^k]$ of order k of the source $[S]$. The relation [3.23] then becomes:

$$\frac{H(S^k)}{\log_2 q} \leq \bar{l}_k < \frac{H(S^k)}{\log_2 q} + 1 \qquad [3.24]$$

Or, according to the relationship [3.22]:

$$\frac{H(S)}{\log_2 q} \leq \frac{\bar{l}_k}{k} < \frac{H(S)}{\log_2 \mathrm{q}} + \frac{1}{k} \qquad [3.25]$$

At the limit, if k tends to infinity, we get:

$$\lim_{k \to \infty} \left(\frac{\bar{l}_k}{k} \right) = \overline{l_{min}} + \varepsilon_k \qquad [3.26]$$

The increased efficiency of the code is performed to the detriment of a greater complexity in coding and decoding the extension $[S^k]$.

NOTE.– Shannon's first theorem is also valid for any stationary source with memory.

In practice, k is finite, so the efficiency is therefore lower than its optimum and the coding is therefore not an optimal coding.

Moreover, to simplify the coding operations, to the detriment of the efficiency, if one chooses $k = 1$, the coding is performed symbol by symbol.

3.9. Design of optimal binary codes

Absolute optimal codes are such that: $l_i = -\log_2 p(s_i)$. Such a condition is usually not verified. The optimal Shannon–Fano and Huffman codings try to best approach this condition.

3.9.1. *Shannon–Fano coding*

The coding algorithm is as follows:

1) we classify the symbols in order of decreasing probabilities;

2) we divide the set of symbols into two subsets of probabilities as close as possible;

3) each subset is prefixed with a bit "0" for one and "1" for the other;

4) we subdivide each subset into two subsets of probabilities that are as close as possible, and do so until the subsets in question only contain one element.

If the different probabilities verify the relationship $p(s_i) = p(c_i) = 2^{-l_i}$ (binary case $q = 2$), then the code obtained is absolute optimal $\overline{l} = \overline{l_{min}}$.

3.9.1.1. *Example of Shannon–Fano binary coding*

Symbol s_k	$p(s_k)$	Partitions					Code C_k
s_1	0.40	0.55 "0"	0.40 "0"				00
s_2	0.15		0.15 "1"				01
s_3	0.15	0.45 "1"	0.25 "0"	0.15 "0"			100
s_4	0.10			0.10 "1"			101
s_5	0.10		0.20 "1"	0.10 "0"			110
s_6	0.06			0.10 "1"	0.06 "0"		1110
s_7	0.02				0.04 "1"	0.02 "0"	11110
s_8	0.02					0.02 "1"	11111

Table 3.4. *Binary coding example of Shannon–Fano coding*

One has:

$$H(S) = 2.483 \text{ bit of information/symbol}$$

$$\overline{l} = 2.59 \text{ bit/codeword}$$

3.9.2. *Huffman code*

The Huffman coding method is based on the application of the following properties. The symbol with the highest probability is assigned to the shortest codeword:

- if $p(s_i) > p(s_j) \Rightarrow l_i \leq l_j$;

– the lengths of the two least likely codewords are equal: $l_{N-1} = l_N$;

– among the codewords of maximum length l_N, there are at least two that are written with the same first l_{N-1} letters.

The coding algorithm takes place in two major phases. The *first major phase* is to iteratively group the two events or groups of events with the smallest probabilities. It breaks down into *two steps* that are carried out iteratively:

– the set of symbols is ordered in order of decreasing probabilities;

– one proceeds to a reduction, which consists of adding the probabilities of the two last symbols of the list $p(s_N) + p(s_{N-1}) = p(r_1)$ and to reorder the $N - 2$ symbols remaining with r_1 $(r_1 = s_N \cup s_{N-1})$. This operation is repeated until we arrive at a sequence of two elements only: $r_n \; r_{n-1}$.

The *second major phase* consists of iteratively constructing the Huffman code by adding a suffix to each of the two events or groups of events with the highest probabilities, a "0" bit for the one and a "1" bit for the other.

If the probabilities of the sets obtained by division are equal, then one would realize an absolute optimal coding.

The following Huffman coding example shows the different steps to build the code.

3.9.2.1. *Example of binary Huffman code designing*

Symbol s_k	$p(s_k)_0$	$p(s_k)_1$	$p(s_k)_2$	$p(s_k)_3$	$p(s_k)_4$	$p(s_k)_5$	$p(s_k)_6$	Code C_k
s_1	0.40	0.40	0.40	0.40	0.40	0.40	0.60 0	1
s_2	0.15	0.15	0.15	0.20	0.25	0.35 0	0.40 1	0 0 1
s_3	0.15	0.15	0.15	0.15	0.20 0	0.25 1		0 1 0
s_4	0.10	0.10	0.10	0.15 0	0.15 1			0 1 1
s_5	0.10	0.10	0.10 0	0.10 1				0 0 0 0
s_6	0.06	0.06 0	0.10 1					0 0 0 1 0
s_7	0.02 0	0.04 1						0 0 0 1 1 0
s_8	0.02 1							0 0 0 1 1 1

Table 3.5. *Example of Huffman binary code designing*

Performance comparison of coding methods:

$$H(S) = -\sum_{i=1}^{8} p(s_i)\log_2 p(s_i) = 2.483 \text{ bit of information/symbol}$$

$$\bar{l} = \sum_{i=1}^{8} p(s_i) \times l_i$$

	Direct coding	Shannon–Fano coding	Huffman coding
$\bar{l}$ bit/symbol	3	2.59	2.54
η	82.7%	95.8%	97.7%
ρ	17.3%	4.2%	2.3%

Table 3.6. *Performance comparison of coding methods*

The Huffman code is the most efficient.

4

Channel Coding for Disturbed Transmission Channels

4.1. Introduction

In Chapter 3, on source coding for non-disturbing transmission channels, a primary source is transformed into a secondary source with an entropy close to its maximum value (very low redundancy), thus making it possible to obtain a quasi maximal efficiency.

In disturbed channels, noise or disturbances in a more general sense, is reflected by the occurrence of discrete errors in the symbols received. To fight against this phenomenon, special redundancy (excess of information) is introduced which one expects to compensate for the ones that the degradations brought about by the transmission channel cause to lose. To this end, to the information symbols, one adds some control symbols whose role is to allow detection or even correction on receipt of transmission errors.

A simplified explanatory block diagram is presented in Figure 4.1 hereinafter, in which the modulator and the demodulator are grouped together with the transmission channel and t here means transposed.

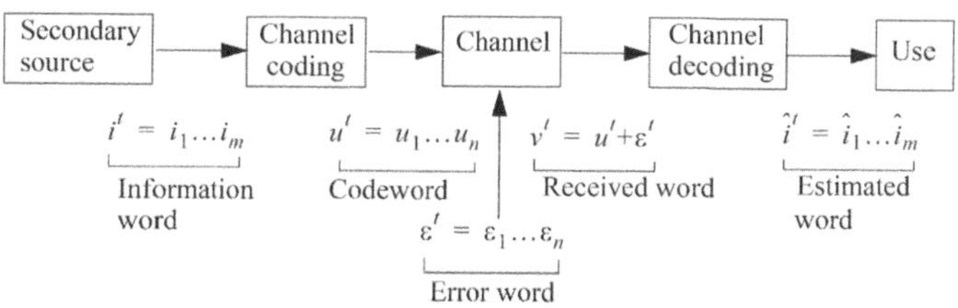

Figure 4.1. *Simplified block diagram of the communication system*

4.2. Shannon's second theorem (1948)

Shannon's theorem for disturbance channels, or Shannon's second theorem, establishes a property whose significance is very important. For a data rate source D (bit/s) and a capacity channel C (bit/s), if $D < C$, there are codes, having codewords of length l, which ensure transmission of symbols with an arbitrarily small probability P_e of decoding error:

$$P_e \leq 2^{-l \times e(D)} \quad [4.1]$$

$e(D)$ is a non-negative function (Figure 4.2) called "exponent of the error probability".

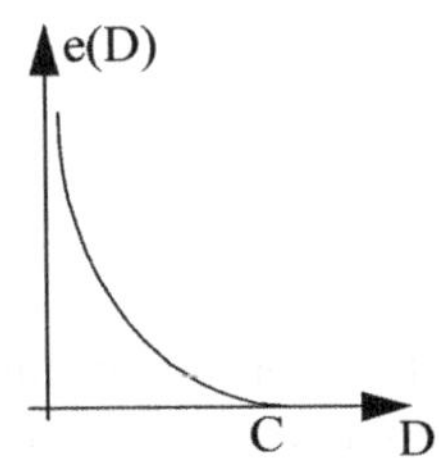

Figure 4.2. *Exponent of the error probability*

However, Shannon's second theorem does not explain how these codes can be designed.

4.3. Error correction strategies

Depending on the nature of the transmission channel, and in general, based on cost considerations, various strategies can be envisaged for detecting or correcting errors.

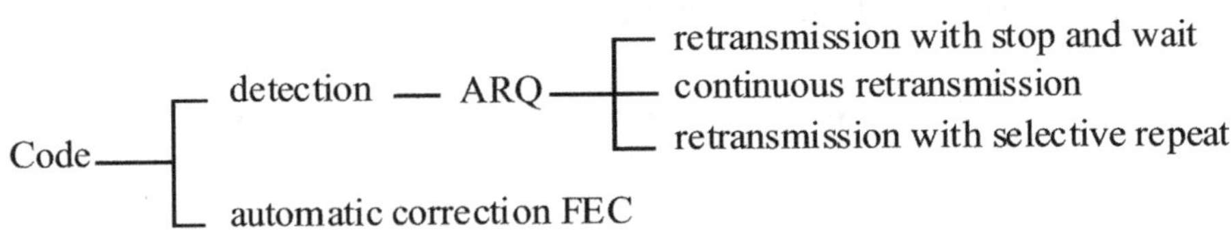

Figure 4.3. *Error correction strategies*

If the channel is bidirectional, error correction can be carried out by retransmission (Automatic Repeat reQuest, ARQ). The least expensive, but also the slowest, is the Stop and Wait retransmission: one transmits a block of information, then waits for the acknowledgment before transmitting the next block, and so on.

In continuous retransmission (Go-back-N), the blocks of information reach the receiver continuously until the decoder reports an error in a certain block, then the retransmission resumes from the erroneous block.

The retransmission with Selective Repeat requires more sophisticated management. In the event of an error, the retransmission of the erroneous block is resumed, and the receiver will take care of reinserting the corrected block at the right place.

Automatic Forward Error Correction (FEC) does not require a return channel: it is the use of error correcting codes. Such a correction system is used for channels with strong disturbances, in order to avoid too frequent repetition, the request for repetition being made only when the number of errors exceeds the possibilities of the correction system.

4.4. Classification of error detection codes or error correction codes

When the processing required to obtain the properties of detection or correction takes place in blocks of n symbols, it is said that we are dealing with linear block codes. The sequences of n symbols are called words.

When the symbols generated by the source are not processed by blocks but continuously, one says that we are dealing with convolutional (or recurrent) codes, not presented in this book.

For linear block codes, we will describe two types of codes, namely group codes and cyclic codes:

– group codes are those in which the words are considered to be elements in a vector space, namely vectors;

– cyclic codes are those in which words are considered to be elements in an algebra, namely polynomials.

4.5. Definitions related to code performance

For the sake of clarity of presentation but also for practical considerations, we will limit ourselves to binary codes ($q = 2$). We then use vector spaces built on the Galois field with two elements, denoted by GF (2).

The two composition laws called "addition modulo 2" and "multiplication" in GF (2) are:

$\oplus$	0	1
0	0	1
1	1	0

Addition modulo 2

$\bullet$	0	1
0	0	0
1	0	1

Multiplication

4.5.1. *Efficiency*

Coding efficiency is the ratio of the number of useful symbols to the number of symbols used:

$$\eta = \frac{m}{n} \quad [4.2]$$

with:

$$n = m + k \quad [4.3]$$

where:

- n: codeword length (protected codeword);
- m: information word length (to be protected);
- k: control word length.

4.5.2. *Weight of linear code or Hamming's weight*

The Hamming weight $P(u^t)$ of a word is the number of non-null binary symbols that the word contains. For example: $P(10011) = 3$.

4.5.3. *Hamming distance*

The Hamming distance between two codewords is equal to the number of different symbols of the same rank they have:

$$d(u_i^t, u_j^t) = \sum_{k=1}^{n} (u_{ik} \oplus u_{jk}) = P(u_i^t \oplus u_j^t) \qquad [4.4]$$

The Hamming distance between two words is none other than the *Hamming weight of their sum modulo 2*. Example:

$$d(10111, 11110) = P(01001) = 2$$

The *minimum weight* of a code is the *smallest word weight* that can be found in it (in the Hamming sense), the zero word (0, ···, 0) excepted.

The *minimum distance* (or distance) of a code is the *smallest distance* "*d*" that can be found between all the codewords taken two by two.

4.6. Form of the decision

Suppose that the words u^t applied to the input of the channel are of length n and taken from a set of cardinal N.

The words received v^t at the output of the channel are also of length n.

The decision consists in assigning to each word v_i^t received a word u_i^t emitted. The decision rule is to split, according to a given criterion, the arrival set V into N subsets V_i such that if $v_i^t \in V_i \Rightarrow d(v_i^t) = u_i^t$, with $\cup_i V_i = V$ and $V_i \cap V_j = 0$ if $i \neq j$.

The conditional probability of error, if u_i^t is sent, is:

$$P_e(i) = \sum_{v_k^t \in \{V - V_i\}} Pr\{v_k^t / u_i^t\} \qquad [4.5]$$

The total average probability of error is:

$$P_e = \sum_{i=1}^{N} P_e(i) \times p(u_i^t) \qquad [4.6]$$

Different criteria can be used to determine the decision rule: that of the maximum likelihood, that of the maximum *a posteriori* likelihood, that of the minimum probability of total average error (that which makes P_e given by [4.6] minimum).

The maximum likelihood rule is to maximize the probability $Pr\{v_j^t/u_j^t\}$.

4.6.1. *Maximum* a posteriori *likelihood decoding*

Let us consider the case of a memoryless binary symmetric channel (BSC) and suppose that each codeword is transmitted with the same probability. Furthermore, let $p < 1/2$ the probability that a received symbol is wrong.

Starting from a received word v_i, the probability that it comes from the transmitted word u_i whose distance to v_i is $d(u_{i,}v_i)$, is:

$$Pr(u_i/v_i) = p^{d(u_{i,}v_i)}(1-p)^{n-d(u_{i,}v_i)} \qquad [4.7]$$

And the probability that it comes from the word sent u_j whose distance to v_i is $d(u_j, v_i)$, is:

$$Pr(u_j/v_i) = p^{d(u_{j,}v_i)}(1-p)^{n-d(u_{j,}v_i)} \qquad [4.8]$$

if:

$$d(u_{i,}v_i) < d(u_{j,}v_i)_{i\neq j} \rightarrow Pr(u_i/v_i) > Pr(u_j/v_i) \qquad [4.9]$$

because:

$$\frac{Pr(u_i/v_i)}{Pr(u_j/v_i)} = p^{d(u_{i,}v_i)-d(u_{j,}v_i)}(1-p)^{d(u_{j,}v_i)-d(u_{i,}v_i)}$$
$$= \left[\frac{1-p}{p}\right]^{d(u_{j,}v_i)-d(u_{i,}v_i)} > 1 \qquad [4.10]$$

The codeword closest to the word received is the one which gives the highest conditional probability $Pr(u_i/v_i)$.

Hence the maximum *a posteriori* likelihood decoding algorithm:

– calculate the distances between the received word and each of the codewords;

– search for the minimum of these distances and take the codeword corresponding to this minimum distance.

It can happen that several codewords are at the same minimum distance from v_i. The choice of one of them is then arbitrary. Such a situation is made all the less frequent as the codewords are at mutual distances as large as possible. This is the principle of Hamming codes.

NOTE.– If $p > 1/2$, it is sufficient to "complement to 1" each of the received bits.

4.7. Linear group codes

Linear group codes are codes in which the control bits depend linearly (according to XOR and AND operators) on the information bits:

– information word:

$$i^t = [i_1, i_2, \cdots, i_m] \ ; \ i_j \in B = \{0,1\}$$

– control word:

$$c^t = [c_1, c_2, \cdots, c_k] \ ; \ c_j \in B$$

– codeword:

$$u^t = [u_1, u_2, \cdots, u_n] \ ; \ u_j \in B$$

with: $n = m + k$.

The code will be called $C(n, m)$.

The linear group codes are such that:

$$c^t = f(i^t) \rightarrow u^t = i^t \mid c^t \qquad [4.11]$$

Where: f is a linear function (with XOR and AND operators).

Therefore, for each block of information (information word) i^t of m bits, a block code (codeword) u^t of n bits, corresponds to it.

The set of words i^t of cardinal $M = 2^m$, forms a vector space E_m of dimension m, that is B^m.

The set of blocks $\{u^t\}$ *of* n *bits, in number* $M = 2^m$ *are the codewords* $C(n, m)$. They form a vector subspace E_m of the vector space E_n of dimension n, that is $N = 2^n$.

The sum modulo 2 of two codewords is a codeword: given u_i^t and u_j^t two codewords of the linear code $C(n, m)$, their sum $u_k^t = [u_i^t \oplus u_j^t]$ belongs to the code $C(n, m)$ and the set of all the possible u_k^t is the set of the codewords, the zero codeword included.

4.7.1. *Decoding ball concept: Hamming's theorem*

The possibilities for detecting and correcting a code depend on the minimum distance from the code in question. The definition of distance makes it possible to construct balls $B(u, r)$ around the M codewords.

The ball with center u_k and radius r is the set of words v whose distance to u_k is at most equal to:

$$B(u_k, r) = \{v / d(u_k, v) \leq r\} \qquad [4.12]$$

In a ball of radius r, there are:

$$\sum_{i=0}^{r} C_n^i \qquad [4.13]$$

codewords of length n, with the binomial coefficient:

$$C_n^p = \frac{n!}{p!\,(n-p)!} \qquad [4.14]$$

The partition of the space of arrival V of the 2^n words received is carried out by constituting M distinct balls having for center the words emitted, all the balls having the same r radius.

4.7.1.1. *Ability to detect and correct errors*

The necessary and sufficient condition (NSC) for a binary code $u_1, u_2, \ldots, u_n$ to make possible the correction of at most r errors is that it fulfills:

$$d_{min}(u_i, u_j) \geq 2r + 1 \qquad [4.15]$$

The necessary and sufficient condition (NSC) for a binary code $u_1, u_2, \ldots, u_n$ to correct r errors is that it checks the following property: $\forall\, i, j = 1, \ldots, n$ and $i \neq j$:

$$d_{min}(u_i, u_j) \geq r + 1 \qquad [4.16]$$

4.7.1.2. *Hamming's theorem*

It provides an upper limit on the number M of words that the code can contain.

Since each ball of radius r contains $\sum_{i=0}^{r} C_n^i$ words belonging to the incoming set of words received, the number M of disjoint balls of radius r fulfills the inequality:

$$M = 2^m \leq \frac{2^n}{\sum_{i=0}^{r} C_n^i} \qquad [4.17]$$

When the inequality [4.17] becomes an equality, we then obtain *perfect codes*, for which the M balls of radius r contain all of the 2^n achievable words.

For $r = 1$, the determination of these perfect codes, also called "*Hamming codes*", is immediate, according to [4.17]:

$$2^{n-m} = 2^k = 1 + n \to n = m + k = 2^k - 1 \qquad [4.18]$$

and:

$$m = 2^k - (k + 1) \qquad [4.19]$$

These codes exist, they correct simple errors, and detect double errors.

The number n is a power of two minus one, hence $n = 3, 7, 15, 31, 63,$ etc., the corresponding values of m are: $m = 1, 4, 11, 26, 57,$ etc.

4.7.2. Generating matrix [G] and test matrix [H]

4.7.2.1. *Recall on matrix: transposed matrix and product of two matrices*

We recall below some useful properties concerning matrices.

4.7.2.1.1. Transposed matrix

Let a matrix:

$$[A]_{m,n} = \left[a_{i,j}\right]_{1\le i\le m,\ 1\le j\le n} \in [M]_{m,n}(K)\ ; \text{example: } [A]_{2,3} = \begin{bmatrix} 1 & 2 & 3 \\ 4 & 5 & 6 \end{bmatrix}$$

with scalar coefficients in K which has a ring structure. The transposed matrix of $[A]_{m,n}$ is the matrix noted:

$$[A]^t_{m,n} = \left[a^t_{i,j}\right]_{1\le i\le m,\ 1\le j\le n} = \left[a_{j,i}\right]_{1\le j\le n,\ 1\le i\le m}$$

$$[A]^t_{m,n} \in [M]_{n,m}(K)\ ; \text{example: } [A]^t_{2,3} = \begin{bmatrix} 1 & 4 \\ 2 & 5 \\ 3 & 6 \end{bmatrix}\ ;\ a^t_{3,2} = a_{2,3} = 6$$

The rows of $[A]$ become the columns of $[A]^t$ and the columns of $[A]$ become the rows of $[A]^t$.

4.7.2.1.2. Properties

$$\left[[A]^t_{m,n}\right]^t = [A]_{m,n}\ ;\ \left[[A]+[B]\right]^t = [A]^t + [B]^t\ ;\ \left[\lambda[A]\right]^t = \lambda[A]^t$$

$$\left[[A]\times[B]\right]^t = [B]^t \times [A]^t\ ;\ [[A]^{-1}]^t = [[A]^t]^{-1}$$

4.7.2.1.3. Product of 2 matrices

To make the product $[P] = [A] \times [B]$ possible, the number of columns in $[A]$ must be equal to the number of rows in $[B]$: $p_{i,j} = \sum_k a_{i,k} \times b_{k,j}$, k being the number of columns in $[A]$ or the number of rows in $[B]$.

Hence:

$$[A]_{m,k} \times [B]_{k,n} = [P]_{m,n}$$

4.7.2.2. *Generating matrix [G]*

In a linear code, the control bits are linked linearly to the information bits, so there is a bijection between the information words and the codewords. The subset of the 2^m words of $\vec{\imath}$ forms a vector subspace in E_n which can be described by a generating matrix $[G']_{m,n}$ of rank m, whose m lines are the linearly independent words of the base:

$$[G']_{m,n} = \begin{bmatrix} u^t_1 \\ u^t_2 \\ \cdots \\ u^t_m \end{bmatrix} \qquad [4.20]$$

Any codeword will be a linear combination of the basic words, so we can get a codeword corresponding to an information word by:

$$u^t = i^t \times [G'] \quad [4.21]$$

The matrix $[G']_{m,n}$ is not unique, a new generating matrix $[G]_{m,n}$, can be obtained from $[G']_{m,n}$ by:

– permutations of the rows;

– permutations of the columns;

– addition to a row, of a linear combination of the other rows, etc.

The codewords remain unchanged, but they will no longer correspond to the same information. We get an equivalent code.

For a systematic code, we have:

$$[u^t] = [i^t \mid c^t] \quad [4.22]$$

($i^t \mid c^t$ is the concatenation of i^t and c^t).

Its generating matrix $[G]_{m,n}$ is written:

$$[G]_{m,n} = \left[I_m \mid P_{m,k}\right] \quad [4.23]$$

Where $[I]_m$ is the identity matrix.

The product:

$$i^t \times [G]_{m,n} = i^t \times \left[I_m \mid P_{m,k}\right] = \left[i^t \mid i^t \times P_{m,k}\right] = [i^t \mid c^t] = [u^t] \quad [4.24]$$

generates a systematic and explicit code.

4.7.2.3. *Test matrix [H]*

Two vectors $\overrightarrow{u_1}$ and $\overrightarrow{u_2}$ are orthogonal if their product is zero:

$$\overrightarrow{u_1} \perp \overrightarrow{u_2} \Leftrightarrow \overrightarrow{u_1^t} \times \overrightarrow{u_2} = 0 \rightarrow \sum_{i=1}^{n} u_{1,i} \times u_{2,i} = 0$$

(summation modulo 2).

Let a code $C(n,m)$ defined by its generating matrix $[G]$. A dual code of $C(n,m)$ is the code $D(n,k)$ defined on the vector subspace orthogonal to the set $C(n,m)$. The matrix $[H]_{k,n}$ generating this dual code, where all the words of $D(n,k)$ are orthogonal to all the words of code $C(n,m)$, is such that:

$$[G]_{m,n} \times [H]_{k,n}^t = [0]_{m,k} \qquad [4.25]$$

Note that: $[H]_{k,n}^t$ stands for $\left[[H]_{k,n}\right]^t$.

For a systematic code, one has:

$$[H]_{k,n} = \left[P_{m,k}^t \mid I_k\right] \rightarrow [H]_{k,n}^t = \begin{bmatrix} P_{m,k} \\ -- \\ I_k \end{bmatrix} \qquad [4.26]$$

From relations [4.23] and [4.26], the relation [4.25] turns into:

$$\left[I_m \mid P_{m,k}\right] \times \begin{bmatrix} P_{m,k} \\ -- \\ I_k \end{bmatrix} = [P]_{m,k} \oplus [P]_{m,k} = [0]_{m,k} \qquad [4.27]$$

4.7.3. *Error detection and correction*

Since:

$$[G] \times [H]^t = [0] \text{ and also } [G] \times [H']^t = [0] \qquad [4.28]$$

Then, every word $u^t \in C(n,m)$ is orthogonal to any row vector of $[H]$ or of $[H']$. Thus:

$$u^t[H]^t = [0], \text{and likewise } u^t[H']^t = [0] \qquad [4.29]$$

The matrix $[H']$ is defined by Hamming (see hereinafter).

The relationships [4.29] determine the coder:

$$c^t = f(i^t)$$

Now, let u^t a codeword sent and v^t the corresponding received word such as:

$$v^t = u^t \oplus \varepsilon^t \qquad [4.30]$$

with ε^t an error word (if no error, then $\varepsilon^t = 0^t$).

The product:

$$v^t[H']^t = u^t[H']^t \oplus \varepsilon^t[H']^t = \varepsilon^t[H']^t = s'^t \quad [4.31]$$

gives the syndrome s'^t, indicating the presence or not of an error and possibly its location.

The relation [4.31] tests the membership of the received word to the code:

– if $s'^t = 0^t \Rightarrow v^t = u^t$, the received word has not been subject to an error or this one is not detectable;

– if $s'^t \neq 0^t \Rightarrow v^t \neq u^t$, the received word has undergone alterations during the transmission.

If the received word v^t contains only one error ($r = 1$), then the relation [4.31] becomes directly exploitable, because in this case:

$$\varepsilon^t = [0, \dots 0, \underset{j^{\text{th}}\ position}{1}, 0, \dots, 0] \Rightarrow s'^t = j^{\text{th}} \text{ row of } [H']^t \quad [4.32]$$

Hence, the possibility of locating the error and correcting it. To do this, it suffices to impose that the binary code of the rows represents the decimal equivalent of the position of the erroneous binary symbol.

4.7.4. *Applications: Hamming codes (r = 1)*

The first non-trivial Hamming code is:

$$C(n, m) = C(7, 4)$$

4.7.4.1. *Detailed study*

The matrix $[H']^t_{3,7}$ proposed by Hamming is:

$$[H']^t_{3,7} = \begin{bmatrix} 0 & 0 & 1 \\ 0 & 1 & 0 \\ 0 & 1 & 1 \\ 1 & 0 & 0 \\ 1 & 0 & 1 \\ 1 & 1 & 0 \\ 1 & 1 & 1 \end{bmatrix} \quad [4.33]$$

The following product determines the control bits:

$$u_{7,1}^t \times [H']_{3,7}^t = [0]_{1,3}$$

$$\rightarrow [u_1 \quad u_2 \quad u_3 \quad u_4 \quad u_5 \quad u_6 \quad u_7] \times \begin{bmatrix} 0 & 0 & 1 \\ 0 & 1 & 0 \\ 0 & 1 & 1 \\ 1 & 0 & 0 \\ 1 & 0 & 1 \\ 1 & 1 & 0 \\ 1 & 1 & 1 \end{bmatrix} = [0 \quad 0 \quad 0] \qquad [4.34]$$

Hence, the control bits:

$$(u_5, u_6, u_7) = f(u_1, u_2, u_3, u_4):$$

$$\begin{cases} Eq1: u_4 \oplus u_5 \oplus u_6 \oplus u_7 = 0 \rightarrow Eq1 \oplus Eq2: u_5 = u_2 \oplus u_3 \oplus u_4 \\ Eq2: u_2 \oplus u_3 \oplus u_6 \oplus u_7 = 0 \rightarrow Eq1 \oplus Eq3: u_6 = u_1 \oplus u_3 \oplus u_4 \\ Eq3: u_1 \oplus u_3 \oplus u_5 \oplus u_7 = 0 \rightarrow Eq1 \oplus Eq2 \oplus Eq3: u_7 = u_1 \oplus u_2 \oplus u_4 \end{cases} \qquad [4.35]$$

The relation [4.35] determines the coder and therefore the matrix $[G]$. Indeed, $[G]$ is expressed:

$$[G]_{4,7} = \begin{bmatrix} 1 & 0 & 0 & 0 & | & 0 & 1 & 1 \\ 0 & 1 & 0 & 0 & | & 1 & 0 & 1 \\ 0 & 0 & 1 & 0 & | & 1 & 1 & 0 \\ 0 & 0 & 0 & 1 & | & 1 & 1 & 1 \end{bmatrix} \quad \text{(columns } u_1\ u_2\ u_3\ u_4\ u_5\ u_6\ u_7\text{)} \qquad [4.36]$$

Moreover, $[G]$ is also obtained from $[H]$, itself obtained from $[H']$ by permutation of columns as follows:

$$[H'] \rightarrow [H]_{3,7} = [P_{4,3}^t \mid I_3] \rightarrow [G] = [I_4 \mid P_{4,3}] \qquad [4.37]$$

$$[H'] = \begin{bmatrix} 0 & 0 & 0 & 1 & 1 & 1 & 1 \\ 0 & 1 & 1 & 0 & 0 & 1 & 1 \\ 1 & 0 & 1 & 0 & 1 & 0 & 1 \end{bmatrix}$$

$$\rightarrow [H] = \begin{bmatrix} 0 & 1 & 1 & 1 & | & 1 & 0 & 0 \\ 1 & 0 & 1 & 1 & | & 0 & 1 & 0 \\ 1 & 1 & 0 & 1 & | & 0 & 0 & 1 \end{bmatrix} \rightarrow [G] \qquad [4.38]$$

The matrix $[G]$ makes it possible to explain the code, since:

$$i^t \times [G] = u^t = [i^t \mid c^t] \qquad [4.39]$$

$$[u_1 \quad u_2 \quad u_3 \quad u_4] \times [G] \quad = \quad [u_1 \quad u_2 \quad u_3 \quad u_4 \quad u_5 \quad u_6 \quad u_7]$$

$$\begin{bmatrix} 0 & 0 & 0 & 0 \\ 0 & 0 & 0 & 1 \\ 0 & 0 & 1 & 0 \\ 0 & 0 & 1 & 1 \\ 0 & 1 & 0 & 0 \\ 0 & 1 & 0 & 1 \\ 0 & 1 & 1 & 0 \\ 0 & 1 & 1 & 1 \\ 1 & 0 & 0 & 0 \\ 1 & 0 & 0 & 1 \\ 1 & 0 & 1 & 0 \\ 1 & 0 & 1 & 1 \\ 1 & 1 & 0 & 0 \\ 1 & 1 & 0 & 1 \\ 1 & 1 & 1 & 0 \\ 1 & 1 & 1 & 1 \end{bmatrix} \times \begin{bmatrix} 1 & 0 & 0 & 0 & 0 & 1 & 1 \\ 0 & 1 & 0 & 0 & 1 & 0 & 1 \\ 0 & 0 & 1 & 0 & 1 & 1 & 0 \\ 0 & 0 & 0 & 1 & 1 & 1 & 1 \end{bmatrix} = \begin{bmatrix} 0 & 0 & 0 & 0 & 0 & 0 & 0 \\ 0 & 0 & 0 & 1 & 1 & 1 & 1 \\ 0 & 0 & 1 & 0 & 1 & 1 & 0 \\ 0 & 0 & 1 & 1 & 0 & 0 & 1 \\ 0 & 1 & 0 & 0 & 1 & 0 & 1 \\ 0 & 1 & 0 & 1 & 0 & 1 & 0 \\ 0 & 1 & 1 & 0 & 0 & 1 & 1 \\ 0 & 1 & 1 & 1 & 1 & 0 & 0 \\ 1 & 0 & 0 & 0 & 0 & 1 & 1 \\ 1 & 0 & 0 & 1 & 1 & 0 & 0 \\ 1 & 0 & 1 & 0 & 1 & 0 & 1 \\ 1 & 0 & 1 & 1 & 0 & 1 & 0 \\ 1 & 1 & 0 & 0 & 1 & 1 & 0 \\ 1 & 1 & 0 & 1 & 0 & 0 & 1 \\ 1 & 1 & 1 & 0 & 0 & 0 & 0 \\ 1 & 1 & 1 & 1 & 1 & 1 & 1 \end{bmatrix} \qquad [4.40]$$

4.7.4.2. *Syndrome calculation*

On reception, the syndrome is calculated as follows:

$$v^t \times [H']^t = s'^t$$

$$\rightarrow [v_1 \quad v_2 \quad v_3 \quad v_4 \quad v_5 \quad v_6 \quad v_7] \times \begin{bmatrix} 0 & 0 & 1 \\ 0 & 1 & 0 \\ 0 & 1 & 1 \\ 1 & 0 & 0 \\ 1 & 0 & 1 \\ 1 & 1 & 0 \\ 1 & 1 & 1 \end{bmatrix} = [s'_3 \quad s'_2 \quad s'_1] \qquad [4.41]$$

$$\rightarrow \begin{cases} s'_3 = v_4 \oplus v_5 \oplus v_6 \oplus v_7 \\ s'_2 = v_2 \oplus v_3 \oplus v_6 \oplus v_7 \\ s'_1 = v_1 \oplus v_3 \oplus v_5 \oplus v_7 \end{cases} \qquad [4.42]$$

EXAMPLE.– Let the information word i^t = [1 1 0 1], the corresponding codeword u^t = [1 1 0 1 | 0 0 1] is obtained by using the relation [4.35] or [4.40]. Let us suppose that one bit has been corrupted during the transmission of u^t, the fourth bit, then the received word will be v^t = [1 1 0 0 | 0 0 1]. One calculates the syndrome with the relation [4.42]. It results in s'^t = [1 0 0], effectively indicating the presence of an erroneous bit in the fourth position.

4.7.5. Coding and decoding circuits

4.7.5.1. *Hamming encoder*

The encoder provides the control bits as a function of the information bits. One has to design the system of relations [4.35] using XOR operator.

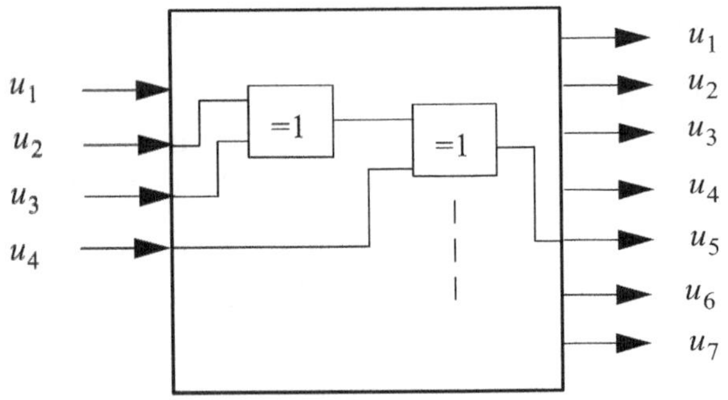

Figure 4.4. *Hamming encoder*

4.7.5.2. *Hamming decoder*

The Hamming decoder carries out the following operations:

– syndrome calculation with [4.42];

– identification of the error position and correction.

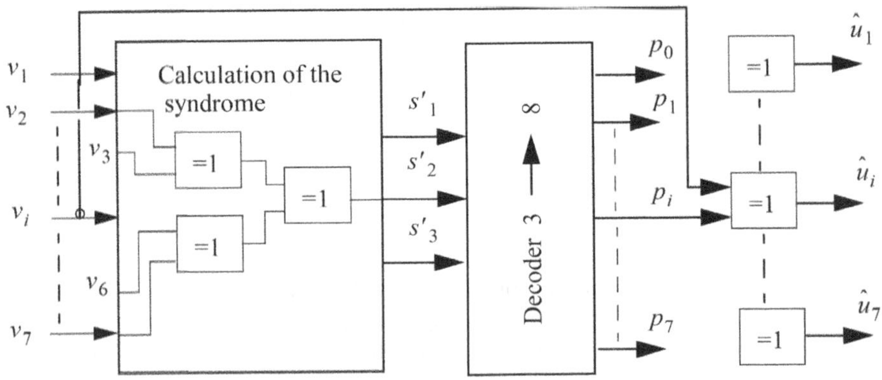

Figure 4.5. *Hamming decoder*

4.7.6. *Extension of Hamming codes*

With the code (7, 4): $d_{min} = 3$, we have the possibility to correct one error at most. This code also makes it possible to detect two errors.

By adding a parity bit to the code $C(7,4)$ on all the bits of the codeword, we get the code $C(8,4)$ of $d_{min} = 4$, which makes it possible to correct one error and to detect two simultaneously.

4.7.6.1. *Extension of the Hamming encoder*

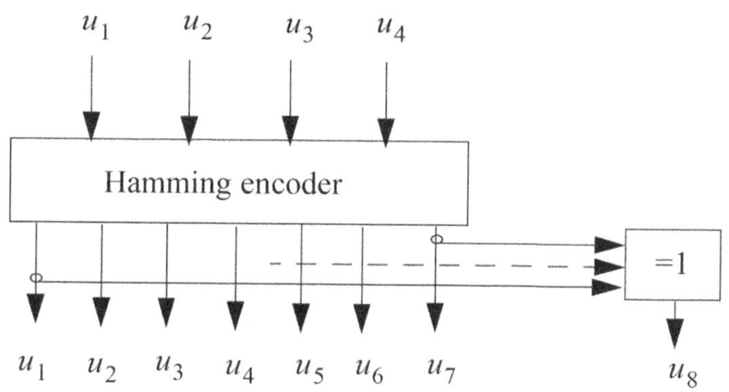

Figure 4.6. *Extension of the Hamming encoder*

4.7.6.2. *Extension of the Hamming decoder*

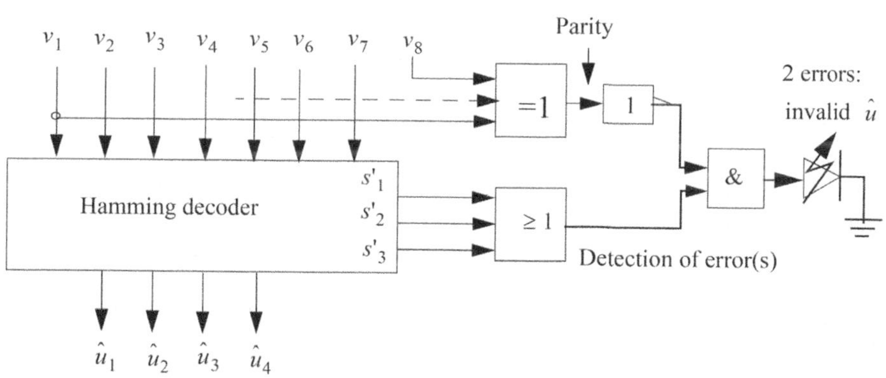

Figure 4.7. *Extension of the Hamming decoder*

From the block diagram we can distinguish the following two cases:

– *simple error = error detection and parity = 1 (odd number of errors);*

– *double error = error detection and parity = 0 (no error or even number of errors).*

NOTE.– Here $n = 2^k$, the code $C(8,4)$ is therefore well suited to protect words in computers, microprocessors and memories. It is designated by SEC DED code (Single Error Correcting, Double Error Detecting).

4.7.7. *Relationships between columns of the matrix [H']*

Let us denote by $h_1^{'}, h_2^{'}, \dots, h_n^{'}$ the columns of the matrix $[H^{'}]$, and assume that the error word ε^t has r errors:

$$\varepsilon^t = \left[\cdots\ 1_{i_1} \cdots 1_{i_r} \cdots\right] \qquad [4.43]$$

where all symbols other than 1 which complete the number of symbols up to n being 0. The indices $i_1, \dots, i_r$ are numbers from 1 to n which show the position in which an error occurs.

The relation $\varepsilon^t[H^{'}]^t = s^{'t}$ is also written:

$$s' = [H']\varepsilon = [h_1' \quad h_2' \quad \cdots \quad h_n'] \times \begin{bmatrix} \vdots \\ 1_{i_1} \\ \vdots \\ 1_{i_r} \\ \vdots \end{bmatrix} = h_{i_1}' \oplus \cdots \oplus h_{i_r}' \qquad [4.44]$$

As the relation [4.44] shows, to be able to correct r errors whatever the positions in which they occur, it is necessary that the sums modulo 2 of each r arbitrary columns of the matrix $[H^{'}]$ are distinct. This will give separate correctors for each error word containing r faults. Otherwise, we will arrive at a situation where r faults occurring in different positions will generate the same syndrome s and, consequently, the correction of the error will no longer be possible.

If the columns of the matrix $[H^{'}]$ are chosen so that the modulo 2 sums of each r columns are distinct, then:

$$h_{i_1}' \oplus \cdots \oplus h_{i_r}' \neq h_{j_1}' \oplus \cdots \oplus h_{j_r}' \qquad [4.45]$$

For all the distinct values $i_1, \ldots, i_r$, likewise for $j_1, \ldots, j_r$ between 1 and n (values that may be equal to some but not all $i_1, \ldots, i_r$ numbers), then:

$$h'_{i_1} \oplus \cdots \oplus h'_{i_r} \oplus h'_{j_1} \oplus \cdots \oplus h'_{j_r} \neq 0 \qquad [4.46]$$

or:

$$h'_{i_1} \oplus \cdots \oplus h'_{i_{2r}} \neq 0 \qquad [4.47]$$

for any value $i_1, \ldots, i_{2r}$ between 1 and n.

It can be shown in the case where the relation [4.47] is fulfilled, that the minimal distance of the code is $2r+1$. Actually, let us suppose that u^t is a codeword of weight equal to $2r+1$, then this codeword could be written:

$$u^t = \left[\cdots\ 1_{i_1} \cdots 1_{i_{2r+1}} \cdots\right] \qquad [4.48]$$

But since u^t is a codeword, we can write the relation like this:

$$[H'] \times u = 0 \ \rightarrow \ h'_{i_1} \oplus \cdots \oplus h'_{i_{2r+1}} = 0 \qquad [4.49]$$

and since the sum of any $2r$ columns is different from zero, according to [4.47], then there is no relation of the form:

$$h'_{i_1} \oplus \cdots \oplus h'_{i_{2r}} = 0 \qquad [4.50]$$

in other words, there will be no word of weight less than $2r+1$ which is the minimum weight, respectively the minimum distance.

4.8. Cyclic codes

4.8.1. *Introduction*

The cyclic codes form a subset of the linear block codes built on the Galois Field GF(2). The construction and use of such codes are very easy, thanks to the algebra of polynomials. The n symbols which constitute a word are considered as the coefficient of a polynomial of degree $n-1$:

$$u^t = [u_{n-1}, u_{n-2}, \cdots, u_i, \cdots, u_1, u_0]$$

$$\Leftrightarrow u(x) = u_{n-1}x^{n-1} + u_{n-2}x^{n-2} + \cdots + u_i x^i + \cdots + u_1 x + u_0 \qquad [4.51]$$

where $u_i \in \{0, 1\}$.

Any left circular permutation performed on the symbols of a codeword gives another codeword:

$$[u_{n-1}, u_{n-2}, \cdots, u_i, \cdots, u_1, u_0] \in C \rightarrow [u_{n-2}, \cdots, u_i, \cdots, u_1, u_0, u_{n-1}] \in C \qquad [4.52]$$

The modulo 2 addition of two codewords is a codeword:

$$\forall\, u_i^t, u_j^t \in C \rightarrow u_i^t \oplus u_j^t \in C \qquad [4.53]$$

The set of all words in number 2^n constitutes an *algebra*, while the set of words in number 2^m having a meaning constitutes an *ideal.*

The interest of these codes is multiple:

– their basic principle relies on the theory of polynomials and algebra;

– they are well suited for detecting independent and packet errors;

– their implementation is easy since the coding and decoding procedures can be made automatic by means of shift registers;

– they are used to check the integrity of the transmitted data received;

– they are used in different layers of transmission network protocols.

Examples of generating polynomials:

– $g(x) = x^8 + x^2 + x + 1$: ATM cell header protection (Asynchronous Transfer Mode);

– $g(x) = x^{16} + x^{12} + x^5 + 1$: CCITT Recommendation No. 41, used in the HDLC transmission protocol (High Data Link Control);

– $g(x) = x^{32} + x^{26} + x^{23} + x^{22} + x^{16} + x^{12} + x^{10} + x^8 + x^7 + x^5 + x^4 + x^2 + 1$: IEEE 802 committee, used in local networks.

4.8.1.1. *Recall of some useful results of algebra*

One defines:

– the *sum of two polynomials*: $u_1(x) \oplus u_2(x) = \sum_{i=0}^{n-1} [u_{1,i} \oplus u_{2,i}]\, x^i$;

– the *product of a polynomial with a Boolean scalar;*

– the *product of two polynomials*: $u(x) = u_1(x) \times u_2(x) \mod x^n + 1$;

– the *Euclidean division of two polynomials*: $u_1(x) = q(x) \times u_2(x) \oplus c(x)$ with: $d°[u_2(x)] \geq 1$, $q(x)$ the quotient polynomial and $c(x)$ the remainder polynomial;

– the modulo $F(x)$ equivalence rule.

The polynomials $u_1(x)$ and $u_2(x)$ are equivalent if the remainders of their division by $F(x)$ are identical.

An equivalence class is represented by its lowest degree polynomial (the remainder polynomial). If $d°[F(x)] = n$, there are 2^n residual classes $\in C_{F(x)}$ (set of residual classes). We are only interested in reducible polynomials $(x^n + 1)$.

Let be the ring of polynomials $u(x)$, and the set of residual classes $C_{F(x)}$ modulo $F(x)$. In $C_{F(x)}$, each ideal is generated by the *lower degree polynomial* $g(x)$ and this element is unique and divides $F(x)$. (An ideal is any part of a closed ring for addition and multiplication, and the set of multiples of an element is a particular ideal).

A polynomial $f(x)$ is *reducible* on a field if there are two polynomials on this field $g(x)$ and $h(x)$ such that: $f(x) = g(x) \times h(x)$. Otherwise, it is *irreducible*.

If $f(x)$ is irreducible, then $C_{F(x)}$ is a field.

4.8.2. *Expression of a circular permutation*

The polynomial representation makes it possible to take advantage of the algebraic structure of the polynomials.

4.8.2.1. *One-step left circular permutation*

Let u^t be a codeword and the corresponding polynomial $u(x)$:

$$[u_{n-1}, u_{n-2}, \cdots, u_i, \cdots, u_1, u_0]$$

$$\Leftrightarrow u_{n-1}x^{n-1} + u_{n-2}x^{n-2} + \cdots + u_i x^i + \cdots + u_1 x + u_0$$

Let us perform a left circular permutation on the symbols of the word and write the corresponding polynomial:

$$[\, u_{n-2}, \cdots, u_i, \cdots, u_1, u_0, u_{n-1} \,]$$

$$\Leftrightarrow u_{n-2}x^{n-1} + \cdots + u_i x^i + \cdots + u_0 x + u_{n-1} \qquad [4.54]$$

It happens as if we had multiplied $u(x)$ by x, but instead of the term $u_{n-1}x^n$ we find the coefficient u_{n-1} multiplying the monomial x^0 instead of x^n. In short, it was stated that:

$$x^n = 1 \text{ or } x^n - 1 = 0 \text{ or again } x^n + 1 = 0 \qquad [4.55]$$

This means to taking $x \times u(x)$ modulo $x^n - 1$ or equivalently, modulo $x^n + 1$, since the addition is performed modulo 2 (XOR).

4.8.2.2. *j-step left circular permutations*

The circular permutation of j positions to the left, corresponds to iterating j times the permutation of one position to the left, and therefore taking the product $x^j \times u(x)$ modulo $x^n + 1$.

EXAMPLE.– Let the word $u^t = [10101]$. A three-step left circular permutation gives $\left[u^{(3)}\right]^t = [01101]$. One finds the same result with the associated polynomials:

$$u(x) = x^4 + x^2 + 1$$

$$u^{(3)}(x) = x^3 \times (x^4 + x^2 + 1) \text{ MOD } (x^5 + 1)$$

One can thus get $u^{(3)}(x)$ by calculating the remainder of the division of $x^7 + x^5 + x^3$ by $x^5 + 1$.

This is equivalent to setting $x^5 = 1$ in $x^7 + x^5 + x^3$. One gets: $x^3 + x^2 + 1$. So one finds the word [01101].

4.8.3. *Generating polynomial g(x), generating matrix [G] and theorem of cyclic codes*

Let us examine the form of the generating matrix of a systematic linear code (relation [4.23]):

$$[G]_{m,n} = \begin{bmatrix} u_1^t \\ u_2^t \\ \vdots \\ u_m^t \end{bmatrix} = \left[I_m \mid P_{m,k}\right] = \left[\begin{array}{ccccc|c} 1 & 0 & \cdots & \cdots & 0 & \\ 0 & 1 & 0 & \cdots & 0 & \\ \cdots & \cdots & \cdots & \cdots & \cdots & P_{m,k} \\ 0 & 0 & 0 & 1 & 0 & \\ 0 & 0 & 0 & 0 & 1 & \end{array}\right] \qquad [4.56]$$

4.8.3.1. *Form of g(x)*

The generating polynomial $g(x)$ of the code corresponds to the codeword (non-zero) of lower degree, therefore:

$$g(x) = g_k x^k + g_{k-1}x^{k-1} + \cdots + g_1 x + g_0 \qquad [4.57]$$

This polynomial is of degree k, in which one has necessarily $g_k = 1$.

On the other hand, $g_0 = 1$, otherwise x could be factorized and the degree of the polynomial would be less than k. So $g(x)$ is of the form:

$$g(x) = x^k + g_{k-1}x^{k-1} + \cdots + g_1 x + 1 \qquad [4.58]$$

4.8.3.2. *Generating matrix [G]$_{m,n}$*

As the polynomial $g(x)$ corresponds to the codeword of lower degree, the circular permutations to the left on the bits of the word g^t are also codewords. Therefore, the polynomials $g(x), xg(x), \ldots, x^{m-1}g(x)$ correspond to m words linearly independent of C, and form a basis for the subspace of the words of C.

Their stacking gives the generating matrix of the code:

$$[G]_{m,n} = \begin{bmatrix} x^{m-1}g(x) & 0 & \cdots & 0 & 0 \\ 0 & x^{m-2}g(x) & 0 & \cdots & 0 \\ \cdots & \cdots & \cdots & \cdots & \cdots \\ 0 & \cdots & 0 & xg(x) & 0 \\ 0 & 0 & \cdots & 0 & g(x) \end{bmatrix} \qquad [4.59]$$

or:

$$[G]_{m,n} = \begin{bmatrix} 1 & g_{k-1} & \cdots & g_1 & 1 & 0 & \cdots & 0 & 0 \\ 0 & 1 & g_{k-1} & \cdots & g_1 & 1 & 0 & \cdots & 0 \\ \cdots & \cdots & \cdots & \cdots & \cdots & \cdots & \cdots & \cdots & \cdots \\ 0 & \cdots & 0 & 1 & g_{k-1} & \cdots & g_1 & 1 & 0 \\ 0 & 0 & \cdots & 0 & 1 & g_{k-1} & \cdots & g_1 & 1 \end{bmatrix} \qquad [4.60]$$

Any codeword is a linear combination of the basic words (the rows of the matrix $[G]$), it is therefore the product of an information polynomial $i(x)$ and of $g(x)$ modulo $x^n + 1$ (see relation [4.21]):

$$u(x) = i(x) \times g(x) \bmod x^n + 1 \qquad [4.61]$$

Thus, any codeword is a multiple of:

$$g(x) \text{ and } d°i(x) = m - 1 \qquad [4.62]$$

The relation [4.61] corresponds to a coding operation by multiplication.

4.8.3.3. *Cyclic codes theorem*

Since the element "0" is a member of an ideal, there exists a polynomial $h(x)$, dual to the polynomial $g(x)$ such that:

$$g(x) \times h(x) = 0 \bmod x^n + 1\,;\; d°h(x) = m = n - k \qquad [4.63]$$

from where:

$$h(x) = \frac{x^n + 1}{g(x)} \rightarrow g(x) \text{ divides } x^n + 1 \qquad [4.64]$$

The degree of $h(x)$ is m: $h(x) = h_m x^m + h_{m-1} x^{m-1} + \cdots h_1 x + h_0$.

Indeed, suppose:

$$x^n + 1 = g(x) \times h(x) + c(x) \text{ with } d°c(x) < d°g(x) = k$$

hence:

$$g(x) \times h(x) + c(x) = 0 \bmod x^n + 1$$

or:

$$g(x) \times h(x) = c(x) \bmod x^n + 1$$

The left side is a codeword (multiple of $g(x)$). The member on the right is necessarily zero, because the non-zero codeword of the lowest degree is precisely $g(x)$, whose degree is k; however, the degree of $c(x)$ is less than k. Hence the *theorem of the cyclic codes* given by relation [4.64] which is thus stated: *a necessary and sufficient condition (NSC) so that a generating polynomial* $g(x)$, *is generator of a cyclic code, is that it divides* $x^n + 1$, *but not* $x^{n1} + 1$ *with* $n1 < n$.

If $C(x)$ is the set of polynomials $u(x)$ associated with the cyclic code C, there are $2^{n-k} = 2^m$ quotients of the divisions of the different polynomials $u(x)$. The size of the cyclic code generated is therefore $2^{n-k} = 2^m$ and it is a code $C(n, m)$.

4.8.4. *Dual code generated by h(x) and parity control matrix [H]*

4.8.4.1. *Dual code generated by h(x)*

The relationship $g(x)h(x) = 0$ modulo $x^n + 1$ recalls the relationship $[G][H]^t = 0$, true for any linear code.

The polynomial $h(x)$ generates a dual code to the code $C(n, m)$. Indeed, let u be a word of a cyclic code generated by the polynomial $g(x)$. The corresponding polynomial $u(x)$ is written: $u(x) = i(x) \times g(x)$.

Let us multiply $u(x)$ by $h(x)$, we get:

$$u(x) \times h(x) = i(x) \times \underbrace{g(x) \times h(x)} = 0 \text{ mod } x^n + 1$$

So:

$$h(x) \times u(x) = 0 \iff [H]u = 0 \qquad [4.65]$$

The polynomial $h(x)$ plays a role similar to that of the matrix $[H]$ and the relation [4.65] makes it possible to determine the control symbols as a function of the information symbols.

4.8.4.2. *Matrix of parity control [H]*

By developing the product $h(x)u(x)$, we show, for all $j \in [1, \ldots, k]$, that:

$$\begin{bmatrix} \underbrace{0\ 0 \cdots 0}_{j-1} \mid h_0\ \ h_1 \cdots h_m \mid \underbrace{0\ 0 \cdots 0}_{k-j} \end{bmatrix} \times \begin{bmatrix} u_{n-1} \\ u_{n-2} \\ \vdots \\ u_0 \end{bmatrix} = 0 \qquad [4.66]$$

By stacking the row-vectors corresponding to the different values of $j \in [1, \ldots, k]$, we obtain the form of the control matrix $[H]$:

$$[H]_{k,n} = \begin{bmatrix} h_0 & h_1 & \cdots & h_m & 0 & \cdots & 0 & 0 \\ 0 & h_0 & h_1 & \cdots & h_m & 0 & \cdots & 0 \\ \cdots & \cdots & \cdots & \cdots & \cdots & \cdots & \cdots & \cdots \\ 0 & \cdots & \cdots & 0 & h_0 & h_1 & \cdots & h_m \end{bmatrix} \qquad [4.67]$$

From [4.66] and [4.67], we can then write:

$$[H]_{k,n} \times [u]_{n,1} = [0]_{k,1} \qquad [4.68]$$

4.8.5. *Construction of the codewords (coding)*

Building a code is to determine the control symbols according to the information symbols. Two methods are possible: coding by division of polynomials and coding by multiplication of polynomials.

4.8.5.1. *Coding by division: systematic code*

Let the information word represented by its polynomial:

$$i(x) = i_{m-1}x^{m-1} + \cdots + i_1 x + i_0$$

The same word can be written:

$$\underbrace{0\,0\,\cdots 0}_{k}\,\underbrace{i_{m-1}\cdots i_1\, i_0}_{m}$$

Let us multiply $i(x)$ by x^k:

$$x^k i(x) = i_{m-1}x^{n-1} + \cdots + i_1 x^{k+1} + i_0 x^k$$

We have just shifted the word $i(x)$ of k positions to the left:

$$\underbrace{i_{m-1}\cdots i_1\, i_0}_{m}\,\underbrace{0\,0\,\cdots 0}_{k}$$

Let us divide $x^k i(x)$ by $g(x)$ (whose degree is k), hence:

$$x^k i(x) = g(x)q(x) + c(x)$$

or:

$$x^k i(x) + c(x) = g(x)q(x) \qquad [4.69]$$

with: $d°c(x) < k$.

The right-hand side of relation [4.69] shows that the word obtained is a member of the code and the left-hand side indicates that this word is composed of two disjointed sub-words: the information word of m bits, concatenated with the control word of k bits:

$$\underbrace{i_{m-1}\cdots i_1\, i_0}_{m}\,\underbrace{c_{k-1}\cdots c_1 c_0}_{k}$$

Thus, the k-bit control word $c(x)$ is obtained by:

$$c(x) = \text{Remainder}\left[\frac{x^k i(x)}{g(x)}\right] \quad [4.70]$$

To summarize, to encode an information word:

– one multiplies it by x^k;

– then, one divides the result by $g(x)$. The remainder of the division gives the control bits.

4.8.5.2. *Coding and decoding circuits: intuitive method through an example*

We describe hereinafter the approach proposed by Pierre Csillag (1990) through an example. It provides the coding and decoding circuits.

Let the following generating and information polynomials:

$$g(x) = x^3 + x + 1 \text{ and } i(x) = x^3 + x$$

$$x^k i(x) = x^3[x^3 + x] = x^6 + x^4 \text{ and } c(x) = \text{Remainder}[(x^6 + x^4)/(x^3 + x + 1)]$$

$$\begin{array}{llcl|ll}
x^k i(x) \longrightarrow & & x^6 + x^4 & & x^3 + x + 1 & \longleftarrow g(x) \\
& - & x^6 + x^4 + x^3 & & x^3 + 1 & \longleftarrow q(x) \\ \cline{3-3}
& & x^3 & & & \\
& - & x^3 + x + 1 & & & \\ \cline{3-3}
& & x + 1 \longleftarrow c(x) & & &
\end{array}$$

Here, the subtraction is without the carry. This is a modulo 2 addition: Exclusive OR (XOR).

With the coefficients only:

$$\begin{array}{rl|ll}
 & 1\,0\,1\,0\,0\,0\,0 & 1\,0\,1\,1 & \longleftarrow g(x) \\
- & 1\,0\,1\,1 & 1\,0\,0\,1 & \longleftarrow q(x) \\
 & 0\,0\,1\,0 & & \\
 & 1\,0\,1\,1 & & \\
 & 0\,1\,0\,0 & & \\
 & 1\,0\,1\,1 & & \\
 & 1\,0\,0\,0 & & \\
- & 1\,0\,1\,1 & & \\
q(x) & 0\,0\,1\,1 \longleftarrow c(x) & &
\end{array}$$

We see the four bits of the divider scrolling from left to right. Subtraction (modulo 2 addition) is performed only if the bit above the left bit of the divider (framed bit in the figure) is 1.

Furthermore, the left bit of the divider (the most significant bit) is always 1 (since the circuit is designed for a divider of a given degree); therefore, the result of the left-most bit subtraction is always 0.

We can therefore imply the most significant bit of the divider and "anticipate" the operations: when the framed bit is 1, the addition will be made as if the three bits of the divider were shifted by one step to the left, and if the framed bit is 0, we will simply copy the bits of the previous result, without adding the bits of the divider:

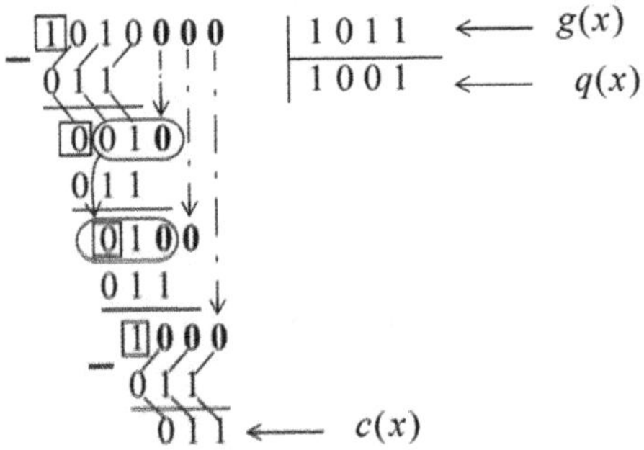

Here, the dividend is fixed, and this is the divider that has been scrolled. It is more practical to fix the divider, which will remain the same, and scroll through the dividend, which will be variable. All this corresponds to a "change of reference" in a kinematic sense, to exchange fixed and mobile references.

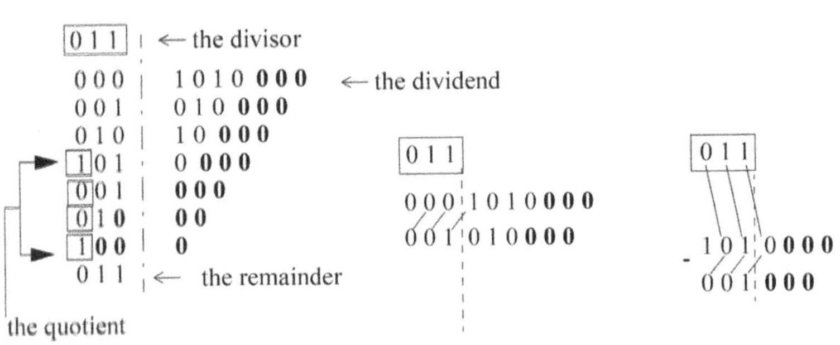

Division diagram

In this case, when a 1 comes to a stop on the left, the subtraction of the bits of the divider will be done again in advance and when a 0 is on the left, the word will only undergo a simple shift. We can therefore materialize this operation by a shift register (based on D flip-flops), in which, each time there is a 1 in the quotient, the dividend will be complemented on the bits corresponding to the non-zero bits of the divisor. As (1) 0 1 1 $\Leftrightarrow g(x) = x^3 + x + 1$, the divider circuit then becomes that shown in Figure 4.8.

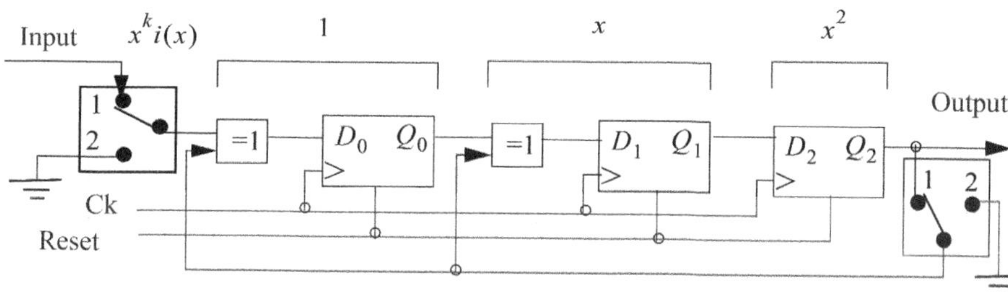

Figure 4.8. *Divider circuit*

In this block diagram:

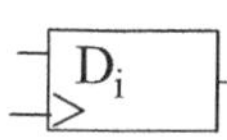

stands for a D flip-flop;

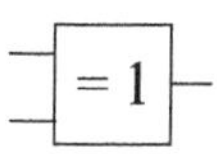

stands for a gate XOR: modulo 2 addition.

OPERATION.–

1) D flip-flops Reset.

2) Switches in position 1: $x^k i(x)$ inputs, most significant bit first during n clock cycles. At the same time k zeros come out then the bits of $q(x)$, most significant bit first. At the end, the bits of the remainder of the division are in the register.

3) Switches in position 2: during k clock cycles, bits $c(x)$ come out, most significant bit first. This circuit uses $n + k$ clock pulses, it is inefficient.

A better encoder can be designed which saves the k over-clocking of the previous coder, by using a divider circuit pre-multiplied by x^k; in other words, a divider in which we no longer need to shift the dividend by k positions. For that, it is enough to convey the input signal k flip-flops further, that is to say, after the last flip-flop on the right. Of course, the input XOR is offset by the same number of positions.

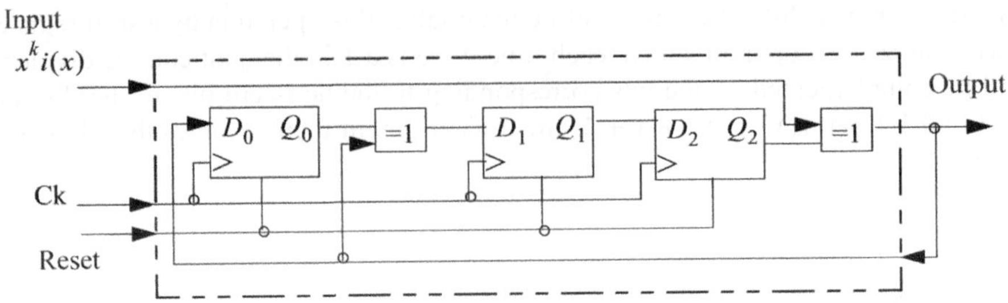

Figure 4.9. *Divider circuit pre-multiplied by x^k*

A coder based on a pre-multiplied divider circuit is shown in Figure 4.10.

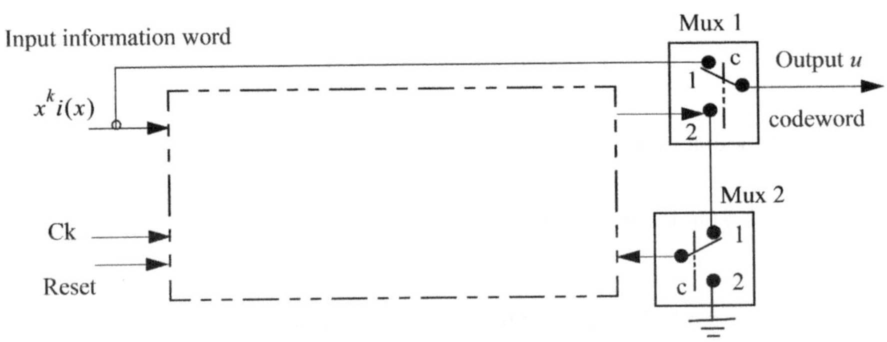

Figure 4.10. *Efficient encoder*

CODING OPERATION.–

1) Reset (clear) all of the flip-flops.

2) During m clock cycles: Mux in position 1: the information bits are applied simultaneously to the divider and to the output. The k control bits are in the k D flip-flops of the register.

3) During k clock cycles: Mux in position 2: zeros enter and the control bits come out.

This encoder uses n clock cycles and the transmission channel is used throughout the operation. It is more efficient than the previous encoder.

EXAMPLE.– Encoder operation.

State of flip-flops and input and output values at each clock cycle:

$$i(x) = x^3 + x\,;\;\; g(x) = x^3 + x + 1$$

$$x^k i(x) = x^6 + x^4 = \; g(x)q(x) + c(x) = [x^3 + x + 1][x^3 + 1] + x + 1$$

$$u(x) = x^k i(x) + c(x) = x^6 + x^4 + x + 1$$

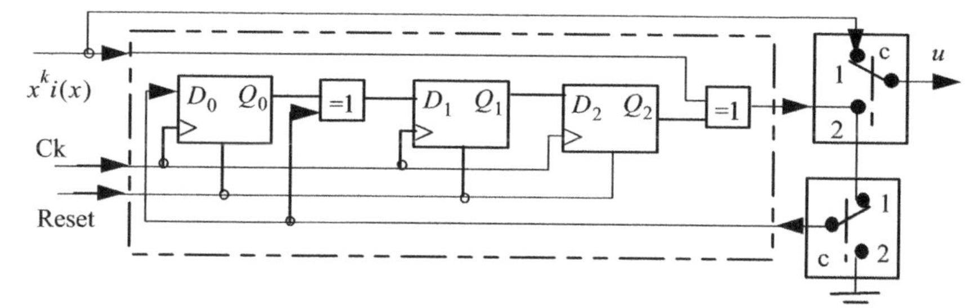

Mu	Ck	$x^k i(x)$	D_0	Q_0	D_1	Q_1	D_2	Q_2		u
1 1		1 x^6	1	**0**	1	**0**	0	**0**		1
	1			1		1		0		
		0 x^5	0		1		1			0
	2			0		1		1	$i(x)$	
		1 x^4	0		0		1			1
	3			0		0		1		
		0 x^3	1		1		0			0
	4			1		1		0	⇐ $c(x)$	
2 2		0 x^2	0		1		1			0
	5			0		1		1		
		0 x^1	0		0		1		$c(x)$	1
	6			0		0		1		
		0 x^0	0		0		0			1
	7			**0**		**0**		**0**	⇐ *Reset*	

Figure 4.11. *Example of encoder operation at each clock cycle*

4.8.5.3. *Transfer function of the division circuit: formal method*

Here we adopt the approach proposed by Alexandru Spataru (1987) to design the coding and decoding circuits. We get the same results as before. Let us divide the polynomial:

$$x^k i(x) = b(x) = b_{n-1}x^{n-1} + b_{n-2}x^{n-2} + \cdots + b_1 x + b_0$$

by the polynomial:

$$g(x) = g_k x^k + g_{k-1}x^{k-1} + \cdots + g_1 x + g_0$$

The following circuit of Figure 4.12 performs this operation.

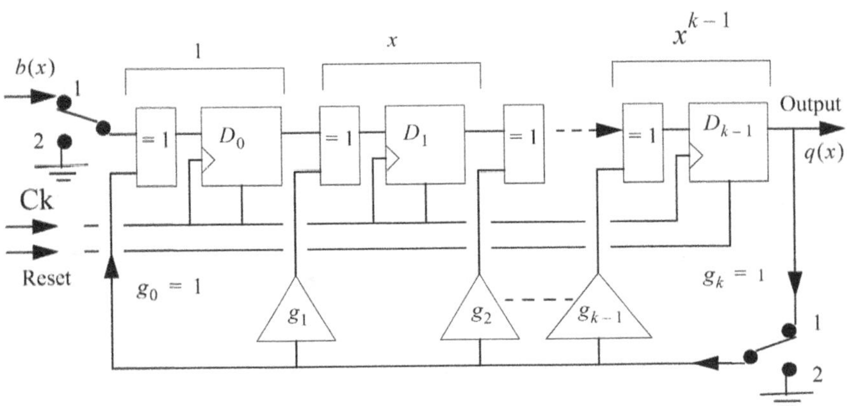

Figure 4.12. *Division circuit by g(x), general case*

In this figure, the coefficients of the polynomial $b(x)$: $b_{n-1}, b_{n-2}, \ldots, b_1, b_0$ enter the division circuit, most significant bit first, during n clock cycles.

Indeed, let $z^{-1} = D$ (Delay), the operator of one clock cycle introduced by a flip-flop cell. The polynomial $b(x)$ is written according to this operator in the form of a binary sequence of data:

$$B(z) = b_{n-1}z^0 + b_{n-2}z^{-1} + \cdots + b_1 z^{-(n-2)} + b_0 z^{-(n-1)} \quad [4.71]$$

The monomial $b_{n-1}z^0$ indicates that the bit b_{n-1} is at the input of the cell D_0 (thus at the entry of flip-flop D_0 because the feedback is zero) and the monomial $b_0 z^{-(n-1)}$ indicates that the bit b_0 comes to the input of the cell D_0 after $(n-1)$ clock cycles.

The transfer function of the division circuit is:

$$T(z) = \frac{Q(z)}{B(z)} \tag{4.72}$$

It is determined as follows. Let us take the case where:

$$B(z) = G(z) = g_k z^0 + g_{k-1}z^{-1} + \cdots + g_1 z^{-(k-1)} + g_0 z^{-k} \tag{4.73}$$

The first symbol g_k applied to the input comes, after k clock cycles to the output of the flip-flop D_{k-1} and therefore, it will appear at the output with a delay z^{-k} so that the first term of the sequence $Q(z)$ will be:

$$Q(z) = g_k z^{-k} = 1 \times z^{-k}$$

So, after k clock cycles, the content of the register is $Q_{i-1} = g_i, i = 1, 2, \dots, k$.

Then, via the feedback connections, we will have:

1) $g_k \times g_0,\ g_k \times g_1, \cdots, g_k \times g_{k-1}$, then;

2) added modulo 2 with the sequence $B(z)$ whose coefficient g_0 is at the input of the cell D_0 and the other coefficients are in the register: $Q_0 = g_1, Q_1 = g_2, \dots, Q_{k-2} = g_{k-1}$. Hence:

$$g_k \times g_0 \oplus g_0,\ g_k \times g_1 \oplus g_1, \cdots, g_k \times g_{k-1} \oplus g_{k-1} = 0$$

As a result of this modulo 2 addition, the symbol 0 will appear at the input of all the flip-flops: $D_i = 0\ \forall\, i = 0, 1, \dots, k-1$.

At the next clock cycle, $k+1$, the content of the register, i.e. $c(x)$ are zero and therefore $1 \times z^{-k}$ will be the only term of the sequence $Q(z)$. In this particular case where $b(x) = g(x)$, the circuit divides $b(x)$ by $g(x)$ with a remainder $c(x)$ null.

The transfer function of the circuit will therefore be:

$$T(z) = \frac{Q(z)}{B(z)} = \frac{z^{-k}}{g_k + g_{k-1}z^{-1} + \cdots + g_1 z^{-(k-1)} + g_0 z^{-k}} \tag{4.74}$$

or:

$$T(z) = \frac{Q(z)}{B(z)} = \frac{1}{g_k z^k + g_{k-1}z^{(k-1)} + \cdots + g_1 z + g_0} \tag{4.75}$$

This result is valid for any polynomial $b(x) = q(x)g(x)$ *divisible by* $g(x)$.

Given $T(z)$, for any data sequence $B(z)$, we can write from [4.75] that:

$$Q(z) = \frac{B(z)}{g_k z^k + g_{k-1} z^{(k-1)} + \cdots + g_1 z + g_0} \quad [4.76]$$

where $B(z)$ can be written from [4.71]:

$$B(z) = z^{-(n-1)}[b_{n-1} z^{n-1} + b_{n-2} z^{n-2} + \cdots + b_1 z + b_0]$$

so:

$$Q(z) = z^{-(n-1)} \times \frac{[b_{n-1} z^{n-1} + b_{n-2} z^{n-2} + \cdots + b_1 z + b_0]}{[g_k z^k + g_{k-1} z^{(k-1)} + \cdots + g_1 z + g_0]} \quad [4.77]$$

Substituting $z = x$ in the relation [4.77], it follows that this circuit will divide the polynomial $b(x)$ by the polynomial $g(x)$. The quotient $q(x)$ of the division will be obtained at the output with a delay $z^{-(n-1)}$, that is after $(n-1)$ clock cycles and the remainder polynomial $c(x)$ of degree $< k$ will then appear stored in the register at the nth clock cycle.

We can output the bits of the remaining polynomial, following the quotient, by k additional clock cycles, entering k zeros both in the circuit input and in the feedback loop.

4.8.5.3.1. Pre-multiplied division circuit by x^k, general case

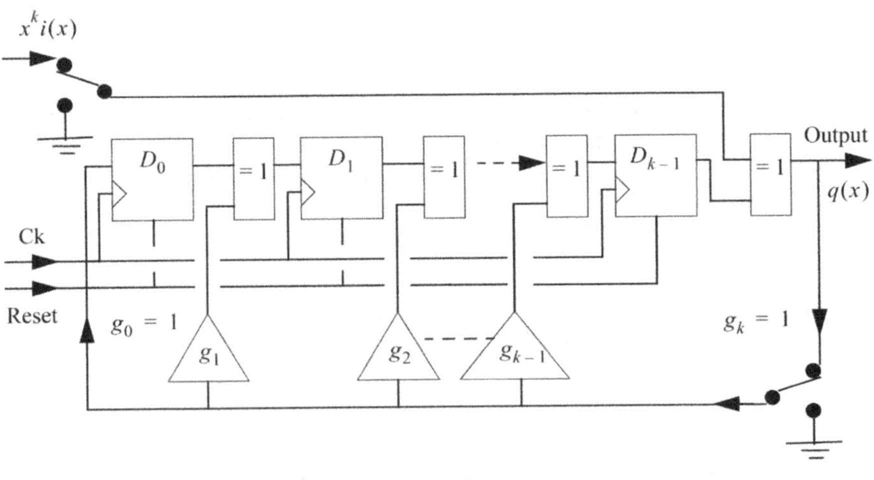

Figure 4.13. *Pre-multiplied division circuit by x^k, general case*

In Figure 4.13, we show the scheme of the pre-multiplied division circuit by x^k in the general case. With this circuit, when the first (most significant) bit of $x^k i(x)$ is introduced, then at the first clock cycle, the contents of the register, initially set to zero, will be $Q_0 = g_0, Q_1 = g_1, \dots, Q_{k-1} = g_{k-1}$. However, k clock cycles were necessary to the dividing circuit in Figure 4.12 to get the same result.

Thus, this pre-multiplied division circuit is more efficient, because it requires in total $n = m + k$ clock cycles:

- m clock cycles to calculate the remainder polynomial;

- k clock cycles to get it out of the register.

4.8.5.4. *Coding and decoding by division circuit, general case*

4.8.5.4.1. Coding

Figure 4.14 shows the block diagram of the encoder based on the previous pre-multiplied division circuit. Its operation is identical to the encoder in Figure 4.10, namely:

1) reset the flip-flops;

2) during m clock cycles: switches c1 and c2, in position 1: the information bits are applied simultaneously to the divider and to the output. The k control bits are in the k flip-flops of the register;

3) during k clock cycles: switches c1 and c2 in position 2, zeros enter and the control bits go out.

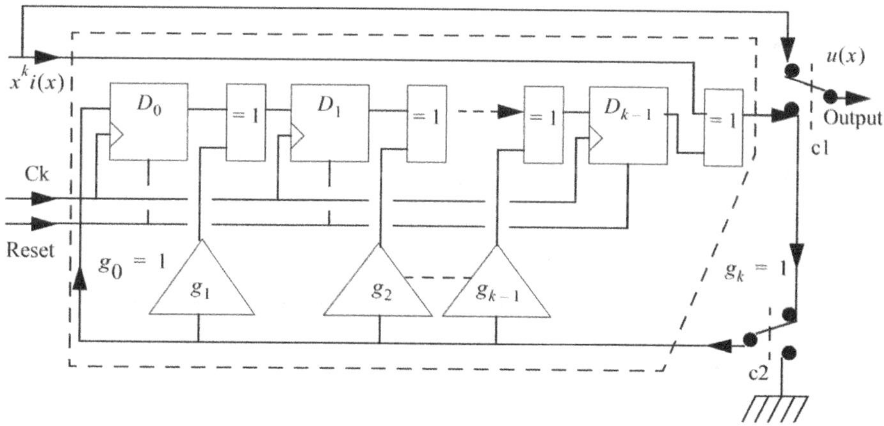

Figure 4.14. *Encoder based on division circuit pre-multiplied by* x^k

4.8.5.4.2. Decoding

Decoding in cyclic codes is used to detect errors and especially error packets in the received word. It is based on the same division circuit used during the coding operation, to calculate the syndrome $s(x)$ which represents the remainder of the division of the received word $v(x)$ by $g(x)$.

In general, the word received $v(x)$ is written:

$$v(x) = u(x) + \varepsilon(x) \quad [4.78]$$

where $u(x)$: codeword sent; $\varepsilon(x)$: possible error word.

The syndrome $s(x)$ is defined by:

$$s(x) = \text{Remainder}\left[\frac{v(x)}{g(x)}\right] = \text{Remainder}\left[\frac{\varepsilon(x)}{g(x)}\right] \quad [4.79]$$

If the syndrome is not zero, this usually means that the received word is wrong, otherwise, the received word is identical to the sent word. However, some configurations of the error words may be divisible by $g(x)$, in this case, despite the presence of the errors, the decoder cannot detect them.

The diagram of the decoder is given in Figure 4.15.

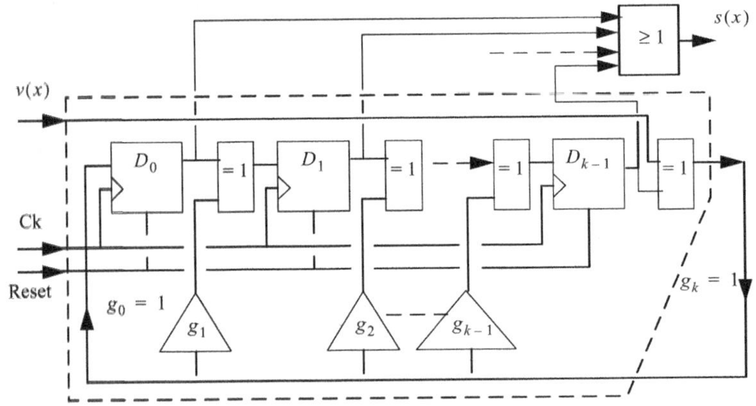

Figure 4.15. *Decoder based on division circuit pre-multiplied by x^k*

OPERATION.– To detect errors, it is sufficient to divide the received word $v(x)$ by the same pre-multiplied division circuit of that of the encoder, for n clock cycles, then to connect the content of the register to the input of a logic gate OR.

4.9. Linear feedback shift register (LFSR) and its applications

The linear feedback shift registers (LFSR) are widely used in digital communications such as:

– data scrambling communications systems;

– spread spectrum for CDMA (Code Division Multiple Access) type applications;

– secure communications using stream-ciphers.

In the rest of this chapter, we will mainly use it:

– in data protection applications transmitted by cyclic coding/decoding based on feedback shift registers;

– in the generation of pseudo-random sequences used in multi-user access applications (CDMA) and for data stream encryption.

The linear feedback shift register is a linear and autonomous sequential circuit based on a feedback signal. Its general scheme is given in Figure 4.16.

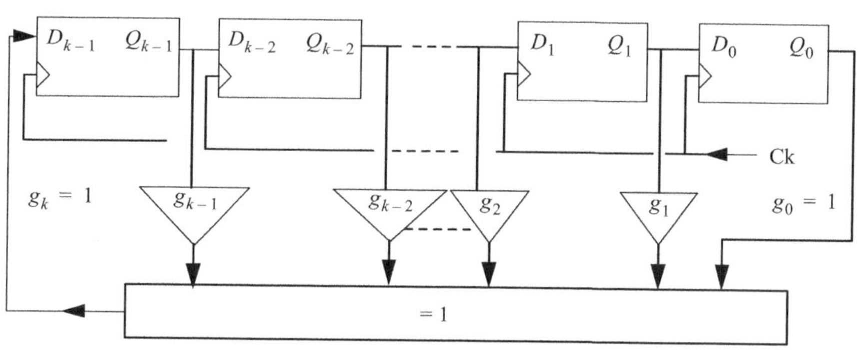

Figure 4.16. *Linear feedback shift register*

The connections are in accordance with the polynomial coefficient:

$$g(x) = g_k x^k + g_{k-1} x^{k-1} + \cdots + g_1 x + g_0$$

OPERATION.– Let us denote by Q_i the state of the flip-flop D_i at time n and by Q_i^+ the state of the same flip-flop D_i at time $n + 1$.

The relationships between the successive states of the register will be:

$$Q_0^+ = Q_1 \,;\; Q_1^+ = Q_2 \,;\; \cdots;\; Q_{k-2}^+ = Q_{k-1} \,;$$

$$Q_{k-1}^{+} = g_0 Q_0 + g_1 Q_1 + \cdots + g_{k-1} Q_{k-1} \quad [4.80]$$

In matrix form:

$$[Q^{+}] = [T][Q] \quad [4.81]$$

with:

$$[T] = \begin{bmatrix} 0 & 1 & 0 & \cdots & \cdots & 0 \\ 0 & 0 & 1 & 0 & \cdots & 0 \\ \cdots & \cdots & \cdots & \cdots & \cdots & \cdots \\ 0 & 0 & \cdots & \cdots & 0 & 1 \\ g_0 & g_1 & \cdots & \cdots & g_{k-2} & g_{k-1} \end{bmatrix} \quad [4.82]$$

and:

$$[Q] = \begin{bmatrix} Q_0 \\ Q_1 \\ \vdots \\ Q_{k-1} \end{bmatrix} \quad [4.83]$$

Where $[T]$ is the characteristic matrix of the register which is invertible as $g_0 = 1$.

4.9.1. *Properties*

If $[Q]$ *is the initial non-zero state of the register*, with for example:

$$[Q]^t = [0, 0, \dots, 1]$$

then the successive states (initial state included) will be:

$$[T]^0[Q], [T][Q], [T]^2[Q], \cdots, [T]^n[Q] = [Q] \quad [4.84]$$

because:

$$n = 2^k - 1 \quad [4.85]$$

is the total number of non-zero states, after which the register returns to its initial state.

We call period of matrix $[T]$, or length of the cycle, the smallest integer n for which:

$$[T]^n = [T]^0 = [I] \qquad [4.86]$$

with $[I]$ the identity matrix.

$2^k - 1$ states can be generated in multiple cycles or in *a single cycle of maximum length*. In the latter case, the generating polynomial $g(x)$ of degree k, is of order $2^k - 1$, it is necessarily *primitive*.

4.9.2. *Linear feedback shift register encoder and decoder (LFSR)*

4.9.2.1. *Linear feedback shift register encoder (LFSR)*

The following encoder of Figure 4.17 implements the following relation:

$$u(x) = x^k i(x) + Remainder[x^k i(x)/g(x)] = x^k i(x) + c(x) \qquad [4.87]$$

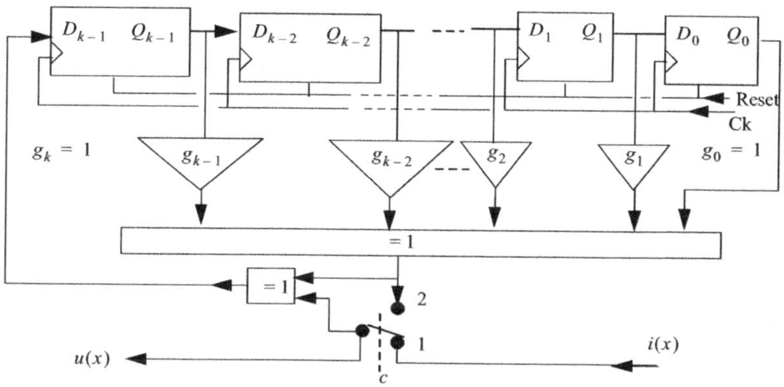

Figure 4.17. *Linear feedback shift register encoder*

OPERATION.–

1) Reset the flip-flops and switch c in position 1.

2) m bits of the information word are entered into the register, most significant bit first. At the same time, they appear at the output:

a) *1st Ck*:

$$Q_{k-1} = u_{n-1}$$

Register state (matrix form):

$$u_{n-1}[Q] \text{ where } [Q]^t = [0 \quad 0 \quad \cdots \quad 0 \quad 1] \quad [4.88]$$

b) *2nd Ck, register state*:

$$u_{n-1}[T][Q] + u_{n-2}[Q] \quad [4.89]$$

c) *mth Ck, register state*:

$$u_{n-1}[T]^{m-1}[Q] + u_{n-2}[T]^{m-2}[Q] + \cdots + u_{n-m}[Q] \quad [4.90]$$

At this time, all the bits of the information word $i(x)$ have been entered into the register and have appeared at the same time as the output. The register contains $c(x)$.

3) Switch c in position 2:

a) *(m+1)th Ck*: a zero enters the register, and the first control bit $u_{n-(m+1)} = c_{k-1}$ goes out. The register state then is:

$$u_{n-1}[T]^m[Q] + u_{n-2}[T]^{m-1}[Q] + \cdots + u_{n-m}[T][Q] + u_{n-(m+1)}[Q] \quad [4.91]$$

And so on, until all k *control bits go out*;

b) *nth Ck, the register returns to its initial state (null)*:

$$u_{n-1}[T]^{n-1}[Q] + u_{n-2}[T]^{n-2}[Q] + \cdots + u_{n-m}[T]^{n-m}[Q]$$

$$+u_{k-1}[T]^{k-1}[Q] + \cdots + u_1[T][Q] + u_0[Q] = [0] \quad [4.92]$$

The relation [4.92] is written in matrix form:

$$[[T]^{n-1}[Q] \quad \cdots \quad [T]^{n-m}[Q] \quad [T]^{k-1}[Q] \quad \cdots \quad [Q]] \times \begin{bmatrix} u_{n-1} \\ \vdots \\ u_{n-m} \\ u_{k-1} \\ \vdots \\ u_0 \end{bmatrix} = [0] \quad [4.93]$$

Note that:

– $[u_{n-1}, u_{n-2}, \ldots, u_{n-m}]^t = [i_{m-1}, i_{m-2}, \ldots, i_0]^t$ are the information bits;

– $[u_{k-1}, u_{k-2}, \ldots, u_0]^t = [c_{k-1}, c_{k-2}, \ldots, c_0]^t$ are the control bits.

The relation [4.93], as well as [4.67], gives the matrix $[H]$ of the cyclic code:

$$[H] = [[T]^{n-1}[Q] \quad \cdots \quad [T]^{n-m}[Q] \quad [T]^{k-1}[Q] \quad \cdots \quad [T][Q] \quad [Q]] \quad [4.94]$$

On the other hand, [4.93] is written as [4.68]:

$$[H][u] = [0] \tag{4.95}$$

4.9.2.2. *Linear feedback shift register decoder (LFSR)*

It is shown in Figure 4.18.

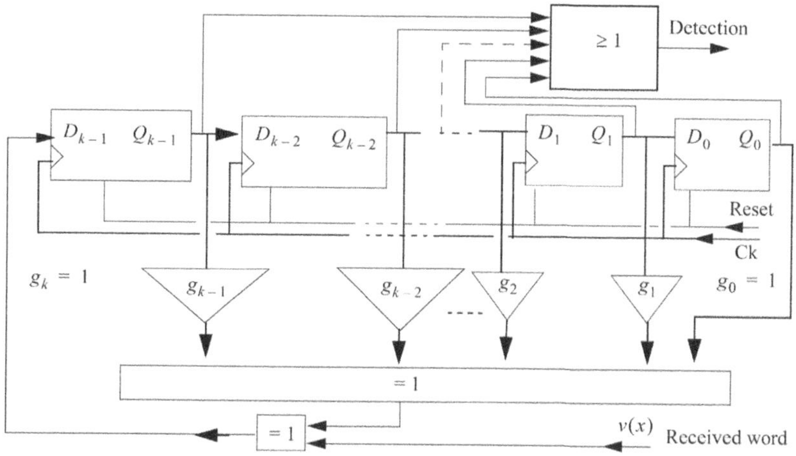

Figure 4.18. *Linear feedback shift register decoder*

OPERATION.– Introduction of the received word $v(x)$ into the register during n clock cycles, the state of the register, according to the reasoning of the previous section, is:

$$v_{n-1}[T]^{n-1}[Q] + v_{n-2}[T]^{n-2}[Q] + \cdots + v_1[T][Q] + v_0[Q] = [s] \tag{4.96}$$

or, in matrix form:

$$[H][v] = [s] \tag{4.97}$$

Verification of the presence of errors is performed by using the relation [4.98]:

$$\text{if } \varepsilon(x)\ \exists \text{ and if } v(x) \notin C(n,m) \Rightarrow s(x) \neq 0 \tag{4.98}$$

EXAMPLE.– Coding and decoding based on LFSR.

1) Make the design of the encoder and decoder based on an LFSR associated with the following primitive generating polynomial:

$$g(x) = x^3 + x + 1$$

2) Determine a) the polynomial $h(x)$, then b) and c) matrix $[H]$ in two different ways.

3) Determine expressions for control bits based on information bits in two different ways.

Initially, the register is initialized to zero.

SOLUTION.– The polynomial:

$$g(x) = x^3 + x + 1$$

is primitive, so:

$$n = 2^k - 1 = 2^3 - 1 = 7$$

and $m = n - k = 4$

1) Encoder and decoder schemes.

The block diagram of the encoder is shown in Figure 4.19; that of the decoder is shown in Figure 4.20.

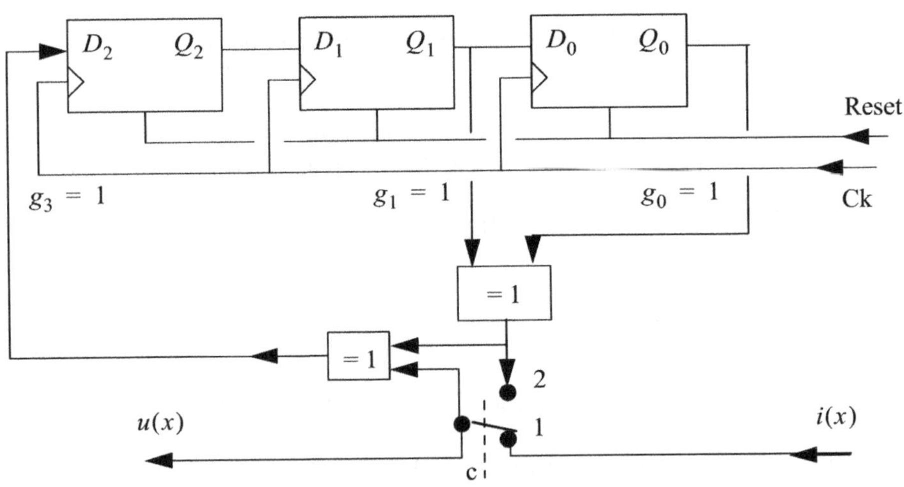

Figure 4.19. *Block diagram of the encoder*

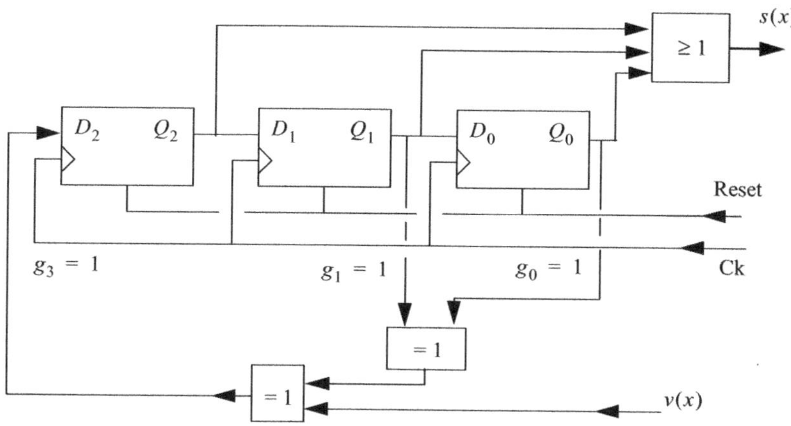

Figure 4.20. *Block diagram of the decoder*

2) Determination of the polynomial $h(x)$, then of the matrix $[H]$:

a) $h(x) = \frac{x^n+1}{g(x)} = \frac{x^7+1}{x^3+x+1} = 1 + x + x^2 + x^4$

b) The form of the matrix $[H]$ of the cyclic code is given by [4.67]:

$$[H]_{k,n} = [H]_{3,7} = \begin{bmatrix} h_0 & h_1 & h_2 & h_3 & h_4 & 0 & 0 \\ 0 & h_0 & h_1 & h_2 & h_3 & h_4 & 0 \\ 0 & 0 & h_0 & h_1 & h_2 & h_3 & h_4 \end{bmatrix}$$
$$= \begin{bmatrix} 1 & 1 & 1 & 0 & 1 & 0 & 0 \\ 0 & 1 & 1 & 1 & 0 & 1 & 0 \\ 0 & 0 & 1 & 1 & 1 & 0 & 1 \end{bmatrix}$$

c) Matrix $[H]$ is also given by [4.94], namely:

$$[H] = [[T]^6[Q] \quad \cdots \quad [T]^3[Q] \quad \cdots \quad [T][Q] \quad [Q]]$$

With, from [4.82] and [4.83]:

$$[T] = \begin{bmatrix} 0 & 1 & 0 \\ 0 & 0 & 1 \\ 1 & 1 & 0 \end{bmatrix}$$

and:

$$[Q] = \begin{bmatrix} Q_0 \\ Q_1 \\ Q_2 \end{bmatrix}$$

State of the register:

– 1st Ck:

$$Q_2 = u_6 = 1 \rightarrow [Q] = \begin{bmatrix} 0 \\ 0 \\ 1 \end{bmatrix}$$

– 2nd Ck:

$$[T][Q] = \begin{bmatrix} 0 & 1 & 0 \\ 0 & 0 & 1 \\ 1 & 1 & 0 \end{bmatrix} \times \begin{bmatrix} 0 \\ 0 \\ 1 \end{bmatrix} = \begin{bmatrix} 0 \\ 1 \\ 0 \end{bmatrix}$$

– 3rd Ck:

$$[T]^2[Q] = [T] \times [T][Q] = \begin{bmatrix} 0 & 1 & 0 \\ 0 & 0 & 1 \\ 1 & 1 & 0 \end{bmatrix} \times \begin{bmatrix} 0 \\ 1 \\ 0 \end{bmatrix} = \begin{bmatrix} 1 \\ 0 \\ 1 \end{bmatrix}$$

– 4th Ck:

$$[T]^3[Q] = [T] \times [T]^2[Q] = \begin{bmatrix} 0 & 1 & 0 \\ 0 & 0 & 1 \\ 1 & 1 & 0 \end{bmatrix} \times \begin{bmatrix} 1 \\ 0 \\ 1 \end{bmatrix} = \begin{bmatrix} 0 \\ 1 \\ 1 \end{bmatrix}$$

– 5th Ck:

$$[T]^4[Q] = [T] \times [T]^3[Q] = \begin{bmatrix} 0 & 1 & 0 \\ 0 & 0 & 1 \\ 1 & 1 & 0 \end{bmatrix} \times \begin{bmatrix} 0 \\ 1 \\ 1 \end{bmatrix} = \begin{bmatrix} 1 \\ 1 \\ 1 \end{bmatrix}$$

– 6th Ck:

$$[T]^5[Q] = [T] \times [T]^4[Q] = \begin{bmatrix} 0 & 1 & 0 \\ 0 & 0 & 1 \\ 1 & 1 & 0 \end{bmatrix} \times \begin{bmatrix} 1 \\ 1 \\ 1 \end{bmatrix} = \begin{bmatrix} 1 \\ 1 \\ 0 \end{bmatrix}$$

– 7th Ck:

$$[T]^6[Q] = [T] \times [T]^5[Q] = \begin{bmatrix} 0 & 1 & 0 \\ 0 & 0 & 1 \\ 1 & 1 & 0 \end{bmatrix} \times \begin{bmatrix} 1 \\ 1 \\ 0 \end{bmatrix} = \begin{bmatrix} 1 \\ 0 \\ 0 \end{bmatrix}$$

Hence, the matrix $[H]$:

$$[H]_{k,n} = [H]_{3,7} = \begin{bmatrix} 1 & 1 & 1 & 0 & 1 & 0 & 0 \\ 0 & 1 & 1 & 1 & 0 & 1 & 0 \\ 0 & 0 & 1 & 1 & 1 & 0 & 1 \end{bmatrix}$$

3) Determination of the expressions of the control bits as a function of the information bits:

a) The relation [4.95], which is identical to [4.68], makes it possible to find the expressions of the control bits as a function of the information bits:

$$[H]_{3,7} \times [u]_{7,1} = [0]_{3,1}$$

so:

$$\begin{bmatrix} 1 & 1 & 1 & 0 & 1 & 0 & 0 \\ 0 & 1 & 1 & 1 & 0 & 1 & 0 \\ 0 & 0 & 1 & 1 & 1 & 0 & 1 \end{bmatrix} \times \begin{bmatrix} u_6 \\ u_5 \\ u_4 \\ u_3 \\ u_2 \\ u_1 \\ u_0 \end{bmatrix} = \begin{bmatrix} 0 \\ 0 \\ 0 \end{bmatrix}$$

Hence:

$$u_6 \oplus u_5 \oplus u_4 \oplus u_2 = 0 \rightarrow \ u_2 = u_4 \oplus u_5 \oplus u_6$$

$$u_5 \oplus u_4 \oplus u_3 \oplus u_1 = 0 \rightarrow \ u_1 = u_3 \oplus u_4 \oplus u_5$$

$$u_4 \oplus u_3 \oplus u_2 \oplus u_0 = 0 \rightarrow \ u_0 = u_2 \oplus u_3 \oplus u_4 = u_3 \oplus u_5 \oplus u_6$$

b) The following coding relation makes it possible to find the same expressions of the control bits as a function of the information bits:

$$c(x) = \text{Remainder}\left\{\frac{x^k i(x)}{g(x)}\right\} = \text{Remainder}\left\{\frac{x^3[u_6x^3 + u_5x^2 + u_4x + u_3]}{x^3 + x + 1}\right\}$$

Divisor: $x^3 + x + 1$

$$\begin{array}{rl} & u_6x^6 + u_5x^5 + u_4x^4 + u_3x^3 \\ \oplus & u_6x^6 + u_6x^4 + u_6x^3 \\ \hline & u_5x^5 + (u_4 + u_6)x^4 + (u_3 + u_6)x^3 \\ & u_5x^5 + u_5x^3 + u_5x^2 \\ \hline & (u_4 + u_6)x^4 + (u_3 + u_5 + u_6)x^3 + u_5x^2 \\ & (u_4 + u_6)x^4 + (u_4 + u_6)x^2 + (u_4 + u_6)x \\ \hline & (u_3 + u_5 + u_6)x^3 + (u_4 + u_5 + u_6)x^2 + (u_4 + u_6)x \\ & (u_3 + u_5 + u_6)x^3 + (u_3 + u_5 + u_6)x + (u_3 + u_5 + u_6) \\ \hline & \underbrace{(u_4 + u_5 + u_6)}_{u_2} x^2 + \underbrace{(u_3 + u_4 + u_5)}_{u_1} x + \underbrace{(u_3 + u_5 + u_6)}_{u_0} \end{array}$$

Quotient: $u_6x^3 + u_5x^2 + (u_4 + u_6)x + (u_3 + u_5 + u_6)$

$$\rightarrow c(x) = u_2x^2 + u_1x + u_0$$

For example, if:

$$i(x) = x^3 + x \to u_6 = 1, u_5 = 0, u_4 = 1, u_3 = 0$$

$$\to u_2 = 0, u_1 = 1, u_0 = 1 \to c(x) = x + 1$$

4.9.3. *Coding by multiplication: non-systematic code*

We know that every word in a cyclic code is a multiple of the generating polynomial:

$$u(x) = i(x)g(x) \text{ mod } x^n + 1 \qquad [4.99]$$

with $i(x)$: word information.

The encoder implements the relation [4.99], indeed:

$$g(x) = g_k x^k + g_{k-1}x^{k-1} + \cdots + g_1 x + g_0 \quad \text{with } g_0 = g_k = 1$$

$$i(x) = i_{m-1}x^{m-1} + \cdots + i_1 x + i_0$$

$$u(x) = \sum_{s=0}^{m-1} i_s x^s \sum_{j=0}^{k} g_j x^j = \sum_{s=0}^{m-1}\sum_{j=0}^{k} i_s\, g_j x^{s+j}$$

Let: $l = s + j$

$$\to u(x) = \sum_{l=0}^{m+k-1}\left[\sum_{s=0}^{m-1} i_s g_{l-s}\right]x^l \quad l - s \in [0, 1, \cdots, k] \qquad [4.100]$$

$$u_l = \left[\sum_{s=0}^{m-1} i_s g_{l-s}\right] \qquad [4.101]$$

The coefficient u_l is the discrete convolution product of the coefficients of the two polynomials.

The expression [4.100] is also written:

$$u(x) = i_0 g_0 + [i_0 g_1 + i_1 g_0]x + [i_0 g_2 + i_1 g_1 + i_2 g_0]x^2 + \cdots + [i_{m-2}g_k + i_{m-1}g_{k-1}]x^{m+k-2} + i_{m-1}g_k x^{m+k-1} \qquad [4.102]$$

For practical implementation, the multiplier $g(x)$ will be fixed and the multiplicand $i(x)$ is variable (input of the encoder).

4.9.3.1. *Practical transposition with an example*

Let:

$$u(x) = i(x) \times g(x)$$

$$u(x) = [x^3 + x^2 + x] \times [x^3 + x + 1] = x^6 + x^5 + x$$

$$[1 \quad 1 \quad 1 \quad 0] \times [1 \quad 0 \quad 1 \quad 1] = [1 \quad 1 \quad 0 \quad 0 \quad 0 \quad 1 \quad 0]$$

Let us make the discrete convolution product degree by degree according to:

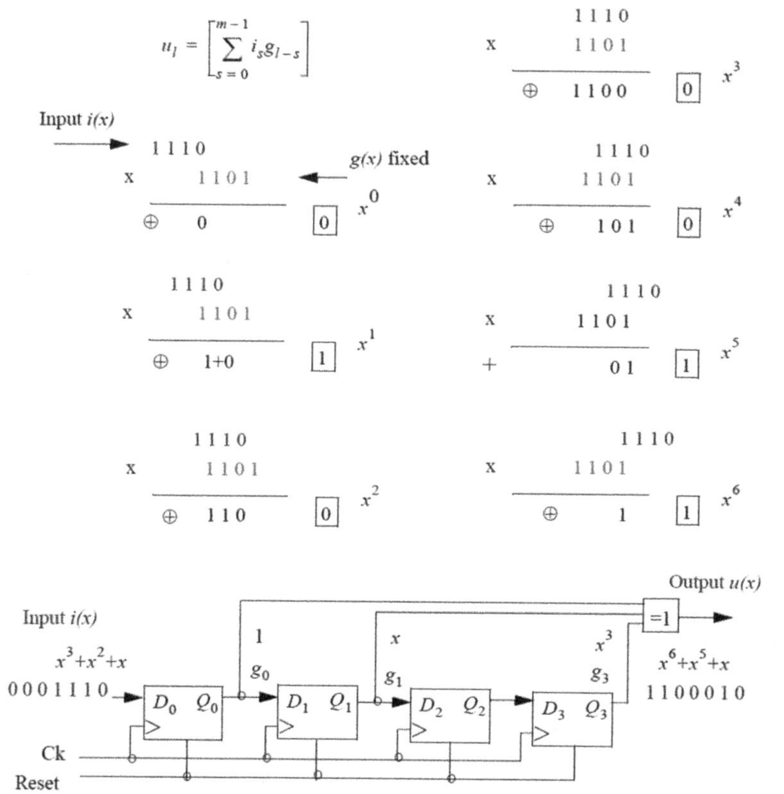

Figure 4.21. *Coding by multiplication. For a color version of this figure, see www.iste.co.uk/assad/digital1.zip*

OPERATION.–

1) Reset of flip-flops.

2) During $m + k = n$ Cks, m bits of information followed by k bits at zero enter the register. During this same time the n bits of the coded word leave the register.

Figure 4.22 shows the state of the register at each clock cycle.

$$u(x) = i(x) \times g(x)$$

$$u(x) = (x^3+x^2+x) \times (x^3+x+1) = x^6+x^5+x$$

$$(1\,1\,1\,0) \times (1\,0\,1\,1) = (1\,1\,0\,0\,0\,1\,0)$$

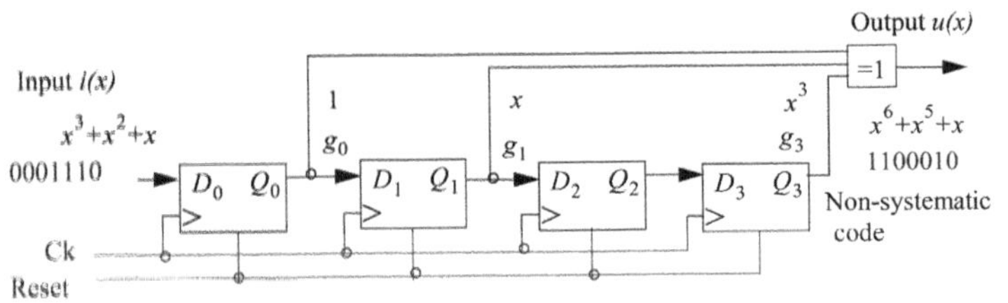

Ck	$i(x)$	D_0	Q_0	D_1	Q_1	D_2	Q_2	D_3	Q_3	u
	0 x^0	0	**0**	0	**0**	0	**0**	0	**0**	
1			0		0		0		0	0
	1 x^1	1		0		0		0		
2			1		0		0		0	1
	1 x^2	1		1		0		0		
3			1		1		0		0	0
	1 x^3	1		1		1		0		
4			1		1		1		0	0
	0 x^4	0		1		1		1		
5			0		1		1		1	0
	0 x^5	0		0		1		1		
6			0		0		1		1	1
	0 x^6	0		0		0		1		
7			0		0		0		1	1
Reset			**0**		**0**		**0**		**0**	u_{n-1}

Figure 4.22. *State of the register at each clock cycle. For a color version of this figure, see www.iste.co.uk/assad/digital1.zip*

4.9.4. *Detection of standard errors with cyclic codes*

The relation $s(x) = \text{Remainder}\{v(x)/g(x)\} = \text{Remainder}\{\varepsilon(x)/g(x)\}$ shows that error detection is possible if $g(x)$ does not divide $\varepsilon(x)$.

One should therefore choose $g(x)$ so that the error patterns $\varepsilon(x)$ we are interested in are not divisible by $g(x)$.

4.9.4.1. *Detection of single error and odd number errors*

A single error results in an error polynomial $\varepsilon(x) = x^i$ which is not divisible by $g(x) = 1 + \cdots$. The lowest degree polynomial $g(x)$ detecting single errors is $g(x) = x + 1$. In this case, the generated cyclic code $\{u(x)\}$ is such that all codewords have an even number of 1.

Indeed, $u(x) = i(x)g(x) = i(x)\,(1 + x)$, hence $u(1) = i(1)(1 + 1) = 0$: simple parity check code $u(1) = u_{n-1} + u_{n-2} + \cdots + u_1 + u_0$.

In addition, $g(x) = x + 1$ not only detects all single errors but also all errors involving an odd number of elements.

This property remains true if $g(x)$ contains $(1 + x)$ in factor:

$$g(x) = (1 + x)(x^{k-1} + \cdots + 1) = (1 + x)p(x) \quad d^\circ p(x) = k - 1 \qquad [4.103]$$

4.9.4.2. *Detection of double errors*

The corresponding polynomial of a double error is $\varepsilon(x) = x^j + x^i = x^i(1 + x^{j-i})$ with $i < j < n$.

Since $g(x)$ does not divide x^i, it is sufficient that $g(x)$ does not divide $(x^{j-i} + 1)$.

The polynomials $g(x)$ are such that $g(x)$ divides $x^n + 1$ but not $x^{n1} + 1$ for $n1 < n$. Then, $g(x)$ is called of order n. If $g(x)$ is of degree k and its order is $n = 2^k - 1$, $g(x)$ is called primitive polynomial (the order $n = 2^k - 1$ is the maximal order of a degree k polynomial). These polynomials are irreducible. They detect all the double errors because $j - i < n$.

4.9.4.3. *Detection of single, double or triple errors*

One has just to choose $g(x) = (1 + x)p(x)$ with $p(x)$ of order n. The single and triple errors are detected by $(1 + x)$. Double errors are detected by $p(x)$.

4.9.4.4. *Error packet detection*

A packet of errors which starts in position j and which has the length l is expressed:

$$\varepsilon(x) = x^j + \varepsilon_{j+1}x^{j+1} + \cdots + x^{l+j-1} = x^j\left[1 + \varepsilon_{j+1}x + \cdots + x^{l-1}\right]$$
$$= x^j \varepsilon 1(x) \qquad [4.104]$$

The first and last coefficients of $\varepsilon 1(x)$ are 1, the others can be either 0 or 1.

Two cases to consider are:

1) $l - 1 < k$: since $g(x)$ is of degree k, it does not divide x^j nor $\varepsilon 1(x)$. Therefore, any cyclic code generated by a polynomial of degree k detects any error packet of length less than or equal to k $(l \leq k)$;

2) $l - 1 \geq k$: the polynomial $\varepsilon 1(x)$ is of degree $l - 1$, but the coefficients of its terms of degree 0 and $l - 1$ are equal to 1, so, there are only 2^{l-2} possible polynomials $\varepsilon 1(x)$. Among these, those that are such that we have:

$$\varepsilon 1(x) = g(x)q(x)$$

lead to undetectable errors.

Two sub-cases can be encountered:

– if $l - 1 = k$ then the degree of $q(x)$ is zero and $q(x) = 1$, hence $\varepsilon 1(x) = g(x)$ and the proportion of undetected error packets of length $l = k + 1$ is then:

$$\frac{1}{2^{l-2}} = 2^{-(k-1)} \qquad [4.105]$$

– if $l - 1 > k$ then the degree of $q(x)$ is $l - 1 - k$. In $q(x)$, the terms of degree 0 and $l - 1 - k$ are not zero, hence there are 2^{l-k-2} possible polynomials such as: $\varepsilon 1(x) = g(x)q(x)$.

The proportion of undetected error packets of length $l > k + 1$ is therefore equal to:

$$\frac{2^{l-k-2}}{2^{l-2}} = 2^{-k} \qquad [4.106]$$

EXAMPLE.– Let $k = 9$; $n = 511$ and $g(x) = x^9 + x^4 + 1$. This code detects single and double errors, packets of errors of length ≤ 9:

– 99.6 % of packets of length $l = 10$;

– 99.8 % of packets of length $l > 10$.

In practice, it is very important to use error packet detecting codes in digital transmission systems because errors occur in bursts rather than in isolation with a uniform probability distribution.

4.9.5. *Pseudo-random sequence generators: M-sequences, Gold, Kasami and Trivium*

Beyond the problem of coding to protect against transmission errors, the security of the transmitted and stored data must be ensured in many applications. Communication security concerns several aspects: the confidentiality of the exchanged information, the integrity of the data against malicious attacks and the authentication of sources.

In what follows, we will only address the problem of confidentiality of the information exchanged or stored. The method used here is stream ciphering. In a stream cipher system, the primary element is the generator of pseudo-random sequences.

4.9.5.1. *Generalities and applications*

Random and pseudo-random sequences can be used:

– in *spread spectrum communication systems* to allow multiple users to use the same transmission channel at the same time;

– in *cryptography* for the security of transmitted and stored data;

– in *generation* of noise.

A random sequence can be obtained from a physical phenomenon having a random character such as:

– an electric gas discharge;

– noise in semiconductor elements, etc.

The random sequence thus obtained after amplification, sampling and quantification, can be recorded and constitutes a secret key used to ensure the secrecy of communications.

DRAWBACK.– Their use in spread spectrum and cryptography applications, for encryption for example, is impossible because they cannot be generated identically on receipt. Hence the use of pseudo-random sequences, because they are:

– reproducible;

– periodicals, period n;

– determinist.

The longer the period n, the closer we can get to the properties of a true random sequence.

4.9.5.2. *Maximum length sequences or M-sequences*

When the connections of a *linear feedback shift register* conform to the coefficients of a primitive polynomial, then the generated sequences are of maximum length $n = 2^k - 1$ called "M-sequences".

Figure 4.23 shows a general block diagram of an M-sequence generator.

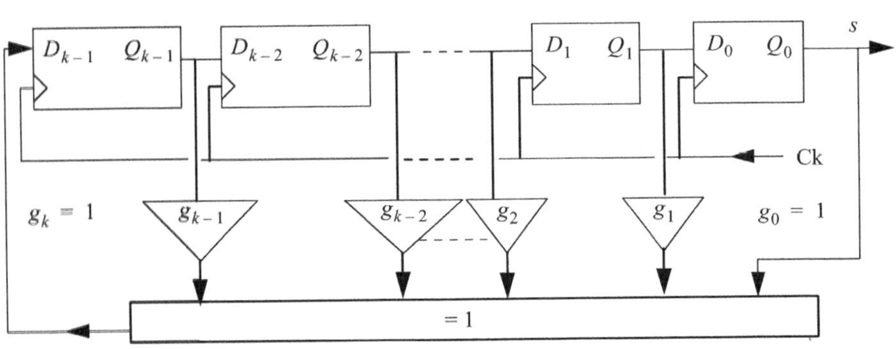

Figure 4.23. *M-sequence generator*

The contents of the register at time n, Q_i^n is changed at time $n+1$ to Q_i^{n+1}.

The register shifts its content to the right at each clock cycle Ck, therefore:

$$Q_i^{n+1} = Q_{i+1}^n \quad i = 0,1,\cdots,k-2 \qquad [4.107]$$

and:

$$Q_{k-1}^{n+1} = g_0 Q_0^n + g_1 Q_1^n + \cdots + g_{k-1} Q_{k-1}^n \quad [4.108]$$

The succession of the state Q_0 of the D_0 flip-flop, represents the sequence s of bits generated by the $Q_0, Q_1, \dots, Q_{k-1}$ register.

In addition, the succession of the states of the other flip-flops represents the same sequence s delivered by Q_0, but shifted:

$$\{Q_i\} = \{Q_{0-i}\}, \; i = 1, 2, \cdots, k-1 \quad [4.109]$$

If $[Q]^t = [Q_0, Q_1, \dots, Q_{k-1}]$ is the non-zero initial state of the register and $[T]$ is its characteristic matrix, then the successive states of the register will be:

$$[Q], [T][Q], [T]^2[Q], \cdots, [T]^{n-1}[Q], [T]^n[Q] = [Q] \quad [4.110]$$

Indeed, consider the following M-sequence generator, constructed from the primitive polynomial $g(x) = x^4 + x^3 + 1$ also denoted [4 3 0], where $k = 4$ and $n = 2^4 - 1$.

Note that T_c is the chip (bit) period.

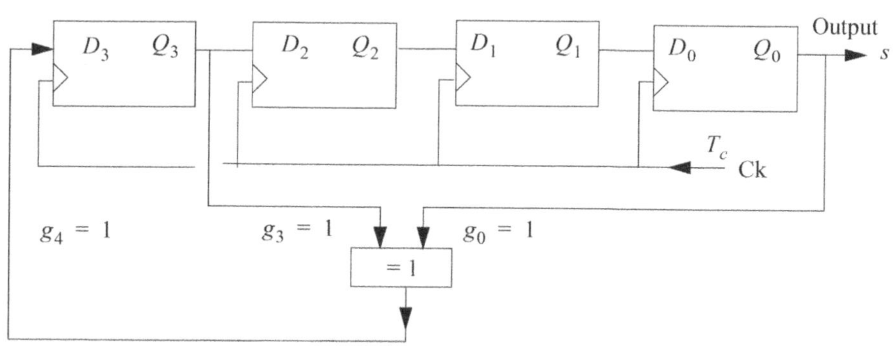

Figure 4.24. *4-cell M-sequence generator*

Let us start from the following initial state:

$$[Q]^t = [Q_0 \quad Q_1 \quad Q_2 \quad Q_3] = [0 \quad 0 \quad 0 \quad 1]$$

The state of the register at each clock cycle Ck is given in Table 4.1.

	Q_3	Q_2	Q_1	Q_0
Ck N°	1	0	0	0
1	1	1	0	0
2	1	1	1	0
3	1	1	1	1
4	0	1	1	1
5	1	0	1	1
6	0	1	0	1
7	1	0	1	0
8	1	1	0	1
9	0	1	1	0
10	0	0	1	1
11	1	0	0	1
12	0	1	0	0
13	0	0	1	0
14	0	0	0	1
15	1	0	0	0

b_0 — Initial state (rows Ck N° and 15)

Table 4.1. *State of the register at each clock cycle*

Figure 4.25 shows the symmetric NRZ signal of the output sequence s.

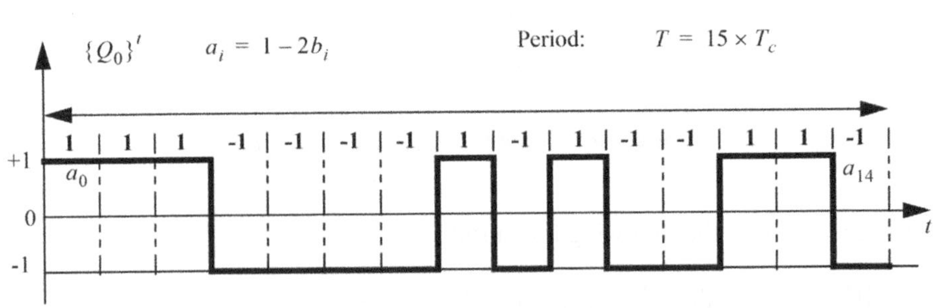

Figure 4.25. *Symmetric NRZ signal of the output sequence (sequence generator of Figure 4.24)*

4.9.5.3. *Properties of M-sequences*

1) Blocking state: $[Q]^t = [0, 0, \dots, 0]$.

2) An M-sequence is quasi well-balanced, it includes:

- $[2^{k-1} - 1]$ zeros, and the probability to have 0 is:

$$Pr\{0\} = \frac{1}{2}\left[1 - \frac{1}{n}\right] \quad [4.111]$$

- 2^{k-1} ones, and the probability to have 1 is:

$$Pr\{1\} = \frac{1}{2}\left[1 + \frac{1}{n}\right] \quad [4.112]$$

There is one 1 more than 0, because the state $[0, 0, \dots, 0]$ is not a member of the set of possible states of the register.

The mean value of the symmetric NRZ sequence is $1/n$, and tends to zero when n is very large.

3) The modulo 2 sum between an M-sequence with an offset version, gives another version of this same M-sequence (version generated by the same generator polynomial, but corresponding to different initial conditions).

4) The statistical distribution of 1s and 0s in a period of an M-sequence is well defined and always the same.

Let us denote by run a series of identical bits of a given length. For example: 000 is a run of "0" of length 3. In a period of the sequence, one has:

– one run of 1 of length k;

– one run of 0 of length $k - 1$;

– 2^{k-r-2} runs of length r, $(r < k)$ of 0 and of 1.

In the example of Figure 4.25, one has:

– one run of 1 of length $k = 4$;

– one run of 0 of length $k - 1 = 3$;

– one run of 1 and one run of 0 of length $r = 2$;

– two runs of 1 and two runs of 0 of length $r = 1$.

The position of these different runs varies from one M-sequence to another, but the number of each of these runs of a particular length is always the same for all M-sequences of the same period n.

5) If a shift register generates an M-sequence, the reciprocal register also generates an M-sequence which is inverted. Let $g(x)$ be a primitive characteristic polynomial, the reciprocal characteristic polynomial which is also primitive is defined by:

$$g_{rec}(x) = x^k g(1/x) \quad [4.113]$$

EXAMPLE.– If $g(x) = x^4 + x^3 + 1$, the corresponding reciprocal polynomial $g_{rec}(x)$ is:

$$g_{rec}(x) = x^4(1 + 1/x^3 + 1/x^4) = x^4 + x + 1$$

We notice on the following M-sequences, that indeed, the M-sequence generated by $g_{rec}(x)$ corresponds to that generated by $g(x)$ but in a reverse sense:

$g(x)$: 0 0 0 1 1 1 1 0 1 0 1 1 0 0 1

$g_{rec}(x)$: 1 0 0 1 1 0 1 0 1 1 1 1 0 0 0

6) The choice of pseudorandom sequences to be used in a spread spectrum system is essentially based on the correlation properties of these sequences.

We recall hereinafter the definitions of the autocorrelation and cross-correlation functions.

4.9.5.3.1. Autocorrelation and cross-correlation functions

The waveform $s(t)$ of a symmetric NRZ M-sequence is expressed as:

$$s(t) = \sum_i a_i p(t - iT_c) \quad [4.114]$$

where:

$$p(t) = \begin{cases} 1 & \text{if } 0 \leq t \leq T_c \\ 0 & \text{elsewhere} \end{cases} \quad [4.115]$$

Since the values of a_i are periodic, of period n, it follows that the signal $s(t)$ is periodic, of period $T = nT_c$.

The autocorrelation of the signal $s(t)$ is:

$$R_s(\tau) = \frac{1}{T}\int_0^T s(t) \times s(t-\tau)\,dt \qquad [4.116]$$

It represents a measure of the resemblance between the signal $s(t)$ and a delayed copy of itself. It is periodic, of period T.

4.9.5.3.2. Cross-correlation function of two waveforms *s*(*t*) and *q*(*t*)

The cross-correlation function of two waveforms $s(t)$ and $q(t) = \sum_i d_i\, p(t - iT_c)$ derived from symmetric NRZ M-sequences is expressed as:

$$R_{sq}(\tau) = \frac{1}{T}\int_0^T s(t) \times q(t-\tau)\,dt \qquad [4.117]$$

It represents a measure of resemblance between the signal $s(t)$ and the signal $q(t)$ shifted in time by τ. The cross-correlation function $R_{sq}(\tau)$ is also periodic with period T.

In the case of discrete M-sequences, the above definitions are transformed hereinafter.

4.9.5.3.3. Discrete, normalized and periodic autocorrelation function

$$R_a(m) = \frac{1}{n}\sum_{i=0}^{n-1} a_i \times a_{i-m} \qquad [4.118]$$

where a_{i-m} is the ith bit (chip) shifted by m positions (m integer) and n as before, the number of bits (chips) per period of the M-sequence.

This autocorrelation function takes only two values:

$$R_a(m) = \begin{cases} 1 \text{ if } m = l \times n \text{ with } l = 0, 1, 2, \cdots \\ \dfrac{-1}{n} \text{ if } m \neq l \times n \end{cases} \qquad [4.119]$$

Indeed, the relation [4.118] is interpreted, by comparing during a period of a M-sequence the bits a_i and a_{i-m} of the same rank, as follows:

$$R_a(m) = \frac{1}{n}[\text{Nb of identical bits } - \text{Nb of different bits}] \qquad [4.120]$$

To illustrate this, let us consider the output sequence Q_2 of the M-sequence generator given in Table 4.1, and a version of the sequence in question, shifted by 2 bits (chips) in the case of symmetric bits (chips).

	a_0														a_{14}
$\{Q_2\}^t =$	1	-1	-1	-1	-1	1	-1	1	-1	-1	1	1	-1	1	1
$\{Q_{2-2}\}^t =$	-1	-1	-1	1	-1	1	-1	-1	1	1	-1	1	1	1	-1
	d	i	i	d	i	i	i	d	d	d	d	i	d	i	d

We get the result given by relations [4.119] and [4.120], namely:

$$R_a(0) = \frac{1}{15}[15] = 1$$

$$R_a(2) = \frac{1}{15}[7-8] = \frac{-1}{15}$$

The autocorrelation function of relation [4.116], after normalization, is shown in Figure 4.26.

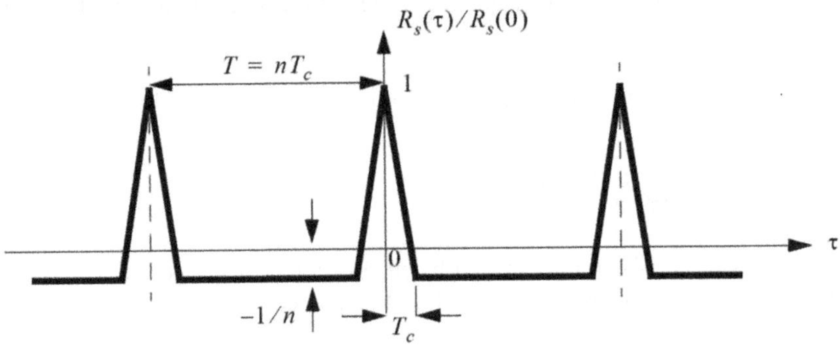

Figure 4.26. *Autocorrelation function of an M-sequence*

Note that for very large n, we can admit that the first peak following that which is at the origin moves away towards infinity, and that the amplitude $-1/n$ between the peaks becomes almost zero, then the M-sequence autocorrelation function converges to that of white noise.

The power spectral density of an M-sequence is expressed in the following relation:

$$\Gamma_s(f) = \frac{1}{n^2}\delta(f) + \frac{n+1}{n^2}\left[\sum_{i=-\infty, i\neq 0}^{\infty}\left[\frac{\sin(i/n)}{i/n}\right]^2 \delta\left(f - \frac{i}{nT_c}\right)\right] \qquad [4.121]$$

It is a discrete spectrum, illustrated in Figure 4.27.

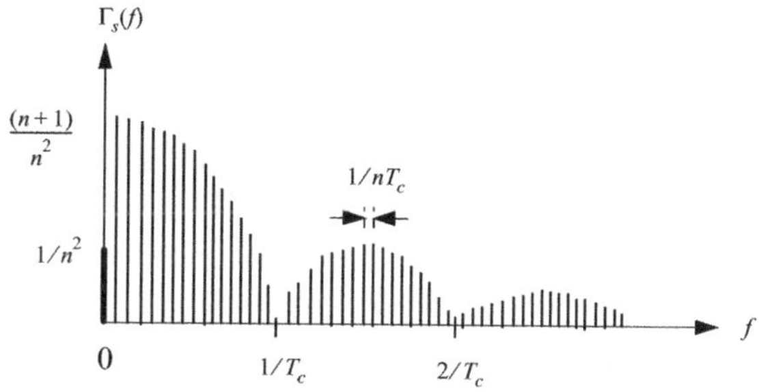

Figure 4.27. *Power spectral density of an M-sequence*

4.9.5.3.4. Discrete, normalized and periodic cross-correlation function

$$R_{ad}(m) = \frac{1}{n}\sum_{i=0}^{n-1} a_i \times d_{i-m} \qquad [4.122]$$

In multi-user systems (CDMA), one should have:

$$\frac{R_{sq\,Max}}{R_{ss}(0)} \ll 1 \qquad [4.123]$$

in such a way that the receiver can synchronize with the appropriate M-sequence.

M-sequences have interesting autocorrelation properties. However, they do not have low cross-correlation values.

The minimum value of the maximum of the cross-correlation of any two M-sequences of period n taken from N M-sequences, of period n each, is given by the Welch bound:

$$R_{sq\,Max} \geq n \times \sqrt{\frac{N-1}{nN-1}} \qquad [4.124]$$

For N and n large, the Welch bound becomes:

$$R_{sq\,Max} \cong \sqrt{n} = \sqrt{2^k - 1} \cong 2^{k/2} \quad [4.125]$$

4.9.5.3.5. Confidentiality with M-sequences

The choice of M-sequences for a sufficiently high degree of confidentiality is not adequate. Indeed, the knowledge of $2k-1$ successive symbols (chips) (with k the number of flip-flops of the register) as well as the period n of the sequence, makes it possible to write $k-1$ equations, which make it possible to find the values of all the coefficients of the generating polynomial $g(x)$ (therefore the feedbacks). The period n can be determined by the precise measurement of the power spectrum of the received signal.

4.9.5.4. *Preferred pairs of M-sequences*

The adequate combination of several specific M-sequences, called "preferred M-sequences", makes it possible to construct pseudo-random generators, such as the Gold generator and that of Kasami. They are capable of producing a number of M-sequences, of length n, greater than the number of M-sequences produced by a simple pseudo-random generator. This increases the number of users in a CDMA communications system. Cross-correlation of pairs of preferred sequences takes only three values which are $\{-1, I(k), I(k)-2\}$ where:

$$I(k) = \begin{cases} 2^{\frac{(k+2)}{2}} + 1 \text{ for } k \text{ even} \\ 2^{\frac{(k+1)}{2}} + 1 \text{ for } k \text{ odd} \end{cases} \quad [4.126]$$

with k equal to the number of flip-flops of each of the two shift registers producing the two original M-sequences.

4.9.5.4.1. Determining whether a pair of M-sequences forms a preferred pair

Let:

– u_j be an M-sequence generated by the irreducible polynomial $g_1(x)$ of degree k and α a root of this polynomial;

– v_j be an M-sequence generated by the irreducible polynomial $g_2(x)$ of degree k.

We show (Penaud 2001) that the sequences u and v form a preferred pair if:

1) k is not divisible by 4: k odd or $k = 2$ modulo 4;

2) $g_2(x)$ is such that:

$\alpha^{2\left[\frac{k-1}{2}\right]+1}$ is a root of $g_2(x)$ if k is odd

$\alpha^{2\left[\frac{k-2}{2}\right]+1}$ is a root of $g_2(x)$ if k is even

EXAMPLE.– Let two polynomials of degree $k = 5$:

$$g_1(x) = x^5 + x^2 + 1: [5, 2, 0]$$

$$g_2(x) = x^5 + x^4 + x^2 + x + 1: [5, 4, 2, 1, 0]$$

If α is a root of $g_1(x)$, then α^5 is a root of $g_2(x)$: check that $g_1(\alpha)$ divides $g_2(\alpha^5)$ or see *Peterson's table*.

Indeed, $g_2(\alpha^5) = \alpha^{25} + \alpha^{20} + \alpha^{10} + \alpha^5 + 1$ can be put in the form:

$$g_2(\alpha^5) = [\alpha^{20} + \alpha^{17} + \alpha^{14} + \alpha^{12} + \alpha^{11} + \alpha^8 + \alpha^7 + \alpha^6 + \alpha^4 + \alpha^2 + 1] \times g_1(\alpha)$$

thus $g_1(\alpha)$ divides $g_2(\alpha^5)$.

4.9.5.5. *Gold sequence generator*

The sequences of Gold (1967) and Kasami (MathWorks 2020) are constructed from pairs of the preferred M-sequences.

These sequences are of length n. They have weaker cross-correlations than M-sequences, as well as a greater number of sequences from a given generator.

Gold has demonstrated that if $g_1(x)$ and $g_2(x)$ are two polynomials generating preferred pairs u and v (M-sequences of length n) then the polynomial:

$$g(x) = g_1(x) \times g_2(x)$$

generates sequences of length n.

One generates a set of $n + 2$ sequences of Gold by:

– addition (module 2) of sequences u and v:

$$G(u, v) = \{u, v, u \oplus D^l v\} \quad l = 0, 1, \cdots, 2^k - 2 \qquad [4.127]$$

where D^l stands for a cyclic shift operator from l positions to the left and $\oplus$ is the "Exclusive OR" logical operator;

– or by using the generating polynomial $g(x)$.

The cross-correlation between the $n+2$ sequences can take only the following three values: $\{-1, I(k), I(k)-2\}$.

EXAMPLE.– Let:

$$g_1(x) = x^5 + x^2 + 1 \text{ and } g_2(x) = x^5 + x^4 + x^2 + x + 1$$

therefore:

$$g(x) = g_1(x) \times g_2(x) = x^{10} + x^9 + x^8 + x^6 + x^5 + x^3 + 1$$

Figure 4.28 shows the block diagram of the Gold generator built from polynomials $g_1(x)$ and $g_2(x)$.

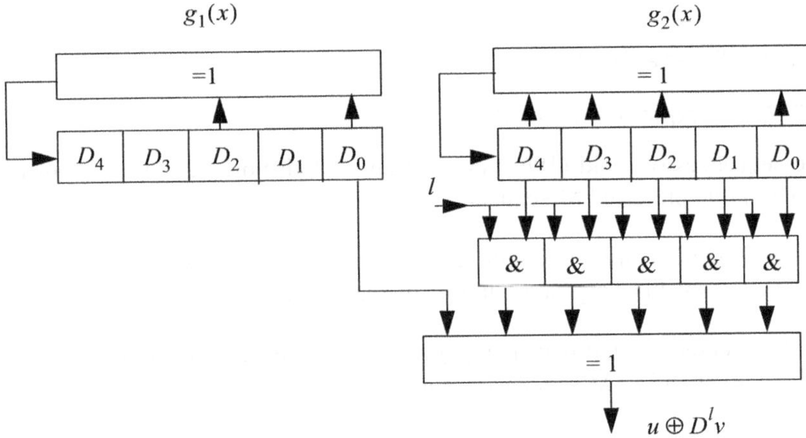

Figure 4.28. *Generator of Gold sequences*

4.9.5.5.1. Summary of the values of the cross-correlation of the M-sequences and the Gold sequences

In Table 4.2 we give, for some values of k, the number of M-sequences, the number of preferred pairs, the number of Gold sequences, the non-normalized and normalized maximum values of the cross-correlation of the M-sequences and Gold sequences.

k	3	4	5	6	7	8	9	10	11	12
$n = 2^k - 1$	7	15	31	63	127	255	511	1,023	2,047	4,095
N_p: number of M-sequences	2	2	6	6	18	16	48	60	176	144
N_{pp}: number of preferred pairs	2	0	3	2	6	–	2	3	4	–
N_{Gold}: number of Gold sequences	9	0	33	65	129	–	513	1,025	2,049	–
$R_{sq\ Max}$	5	9	11	23	41	95	113	383	287	1,047
$R_{sq\ Max}/R_{ss}(0)$	0.71	0.6	0.35	0.36	0.32	0.37	0.22	0.37	0.14	0.34
$I(k)$	5	9	9	17	17	33	33	65	65	129
$I(k)/R_{ss}(0)$	0.71	0.6	0.29	0.27	0.13	0.13	0.06	0.06	0.03	0.03

Table 4.2. *Summary of the values of the cross-correlation of the M-sequences and Gold sequences*

We notice that:

– for $k < 7$, the normalized values of the cross-correlation of the M-sequences and of the Gold sequences are not good for a CDMA type application;

– for $k \geq 7$ and larger k, the normalized cross-correlations of the Gold sequences become better than those of the M-sequences. In addition, the number of Gold sequences becomes large, which is necessary in a multi-user communication system.

Furthermore, the Gold sequences are particularly effective in the synchronization process (Dixon 1994).

4.9.5.6. *Kasami sequence generator*

Like the Gold sequences, the Kasami sequences can be generated in the same way. There are two sets of Kasami sequences; the small set and the large set. The large Kasami set contains the Gold sequences and the small Kasami set.

4.9.5.6.1. Small Kasami set

The small Kasami set of reduced cardinal $2^{k/2}$ is constructed as follows.

Let u an M-sequence, of period $n = 2^k - 1$, with k even. A sequence w is formed by decimation of u by $2^{k/2} + 1$. The sequence w is periodic, of period $2^{k/2} - 1$. A period n of the sequence u, has $2^{k/2} + 1$ periods of the sequence w.

EXAMPLE.– For $k = 6$, the cardinal number is 8, the period of u is $n = 63$, the period of w is $2^3 - 1 = 7$ and in a period of u, there are $2^3 + 1 = 9$ periods of the sequence w.

The Kasami sequences are formed by summing modulo 2, the sequence u and the sequences w obtained by cyclic shift to the left:

$$K_p(u, k, m) = \{u, u \oplus D^m w\} \quad m = 0, 1, \cdots, 2^{k/2} - 2 \qquad [4.128]$$

One thus obtains $2^{k/2} - 1$ different sequences, to which one adds the sequence u to form the small set of Kasami.

The autocorrelation ($R(0)$ excluded) and cross-correlation functions of these sequences take three different values: $\{-1, -[2^{k/2} + 1], 2^{k/2} - 1\}$.

The small set of Kasami is quasi optimal in the sense of the Welch bound as the maximum cross-correlation of two sequences is:

$$R_{sq\,Max} = 2^{k/2} + 1 \qquad [4.129]$$

4.9.5.6.2. Large Kasami set

The large Kasami set consists of sequences of period $n = 2^k - 1$, for k even of the form $k = 4i + 2$, i integer. The sequences of the large Kasami set are generated in the following manner.

From the initial sequence u (M-sequence) of period $n = 2^k - 1$, we form a new sequence v by decimation of the sequence u by $2^{(k+2)/2} + 1$ and a second sequence w by decimation of the sequence u by $2^{k/2} + 1$.

The period of the sequence v is identical to that of the sequence u and the period of the sequence w is $2^{k/2} - 1$. The large Kasami set is formed as follows:

$$K_g(u, k, l, m) = \{u \oplus D^l v \oplus D^m w\} \quad l = 0, 1, \cdots, 2^k - 2; \qquad [4.130]$$
$$m = 0, 1, \cdots, 2^{k/2} - 2$$

For greater clarity, the large Kasami set is also detailed in [4.132].

The cardinal of this set is: $[2^k + 1]2^{k/2}$.

The autocorrelation ($R(0)$ excepted) and cross-correlation functions of these sequences take five different values: $\{-1,[-1\pm 2^{k/2}],[-1\pm 2^{(k+2)/2}]\}$. The maximum value of the cross-correlation functions of two sequences is:

$$R_{sq\,Max} = 1 + 2^{(k+2)/2} \qquad [4.131]$$

It is not optimal in the sense of the Welch bound.

$$K_g(u,k,l,m) = \begin{cases} u\ (\text{M}-\text{Sequence}) & l = 2^k\ ;\ m = 2^{k/2}-1 \\ v\ (\text{M}-\text{Sequence}) & l = 2^k - 1\ ;\ m = 2^{k/2}-1 \\ G(u,v) = u \oplus D^l v & 0 \le l \le 2^k - 2\ ;\ m = 2^{k/2}-1 \\ K_p = u \oplus D^m w & l = 2^k - 1\ ;\ 0 \le m \le 2^{k/2} - 2 \\ v \oplus D^m w & l = 2^k\ ;\ 0 \le m \le 2^{k/2} - 2 \\ K_g = \{u \oplus D^l v \oplus D^m w\} & 0 \le l \le 2^k - 2\ ;\ 0 \le m \le 2^{k/2} - 2 \end{cases} \qquad [4.132]$$

EXAMPLE.– Hereinafter, we give an example of the construction of a large Kasami set for $k = 6$ [Glas]. This set will consist of 520 sequences of period $n = 63$, of which 240 sequences are balanced and therefore usable in a multi-user system. The three generating polynomials used are:

$g_1(x) = x^6 + x + 1$ primitive: generates the sequence u

$g_2(x) = x^6 + x^5 + x^2 + x + 1$ primitive: generates the sequence v

$g_3(x) = x^3 + x^2 + 1$ primitive: generates the sequence w

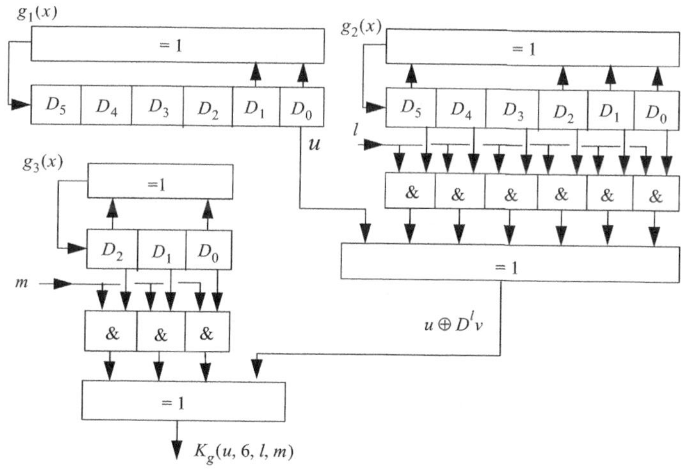

Figure 4.29. *Sequences generator of the large Kasami set*

The sequences u and v form a preferred pair and the sequence w is obtained by decimation of the sequence u by $2^3 + 1 = 9$.

4.9.5.7. *Trivium non-linear generator*

This is one of the standard generators used for stream ciphering (eStream project). The block diagram of a stream ciphering/deciphering system is presented in Figure 4.30 (De Cannière and Preneel 2006; Maximov and Biryukov 2007; Paar and Oelzl 2010).

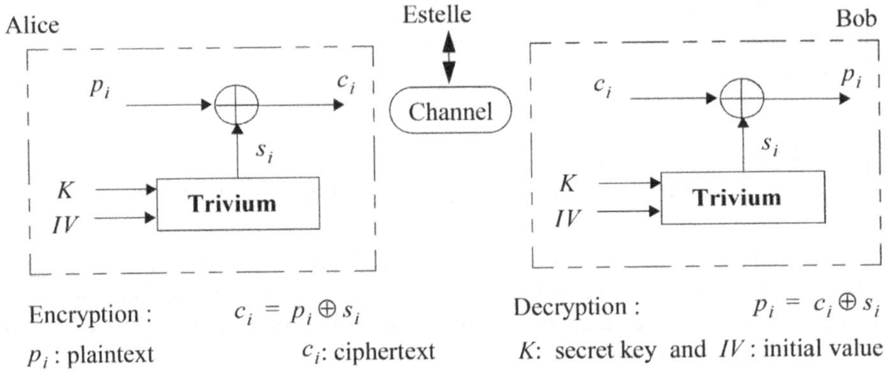

Figure 4.30. *Block diagram of a stream encryption/decryption system*

The stream encryption/decryption algorithm comes down to an XOR operation between:

– the plaintext and the keystream s_i for encryption;

– the ciphertext and the keystream s_i for decryption.

Table 4.3 summarizes this operation.

p_i	s_i	c_i
0	0	0
0	1	1
1	0	1
1	1	0

Table 4.3. *XOR operator*

The security of the encryption algorithm, for a rather random secret key of size greater than or equal to 128 bits, then depends entirely on the sequence s_i delivered by the generator (keystream). Indeed:

– cipher $p_i = 0$ depends on $s_i = \begin{cases} 0 \\ 1 \end{cases} \Rightarrow c_i = \begin{cases} 0 \\ 1 \end{cases}$;

– cipher $p_i = 1$ depends on $s_i = \begin{cases} 0 \\ 1 \end{cases} \Rightarrow c_i = \begin{cases} 1 \\ 0 \end{cases}$.

If the bit stream s_i is perfectly random, then the probabilities:

$$Pr\{s_i = 0\} = Pr\{s_i = 1\} = 0.5$$

Consequently, the probabilities:

$$Pr\{c_i = 0\} = Pr\{c_i = 1\} = 0.5$$

Suppose that the keystream period tends to infinity, in this case the encryption/decryption system is perfect (unconditionally secure). This cipher is called *one-time pad*.

4.9.5.7.1. Need of an initial value (IV) to overcome known plain text attack

Let us suppose that Estelle has a ciphertext and its corresponding plaintext, she can then easily find the keystream $\{s_i\} = \{c_i\} \oplus \{p_i\}$. Moreover, if she has two different ciphertexts $\{c_1\}$ and $\{c_2\}$ encrypted with the same keystream $\{s_i\}$, she can then XOR them together and get the two plain text sequences XORed with each other: $\{c_1\} \oplus \{c_2\} = \{p_1\} \oplus \{p_2\}$. This is easy to break, and then she can XOR one of the plaintexts with its ciphertext to get the keystream. Now, each time she intercepts a new encrypted sequence generated by the same secret key K, she can decrypt it. In addition, she can decrypt any old encrypted text she has previously intercepted.

Therefore, the value of the initial value (IV) must be changed with each new session and must be used only once. Thus, the sequences generated in the different sessions with the same secret key K are different.

4.9.5.7.2. Internal structure of Trivium

The output of each register is connected to the input of another register so that the whole forms a single circular register of length: 93 + 84 + 111 = 288 states, noted $[s_1, s_2, \ldots, s_{288}]$. The AND gates are the only non-linear elements to prevent attack which exploits the linearity of the linear feedback shift registers.

Register	Register length	Feedforward bit	Feedback bit	AND inputs
A	93	66	69	91; 92
B	84	69	78	82; 83
C	111	66	87	109; 110

Table 4.4. *Specifications*

Secret key size *K*	Size of *IV*	Internal state size	Maximum keystream size
80 bits	80 bits	288 bits	2^{64} bits

Table 4.5. *Parameters*

OPERATION.–

– *Initialization*: the initialization phase takes place as described hereinafter:

$$\begin{cases} [s_1, s_2, \ldots, s_{93}] \leftarrow [k_1, k_2, \cdots, k_{80}, 0, \cdots, 0\,] \\ [s_{94}, s_{95}, \ldots, s_{177}] \leftarrow [iv_1, iv_2, \cdots, iv_{80}, 0, \cdots, 0\,] \\ [s_{178}, s_{179}, \ldots, s_{285}, s_{286}, s_{287}, s_{288}\,] \leftarrow [0, 0, \cdots, 0, 1, 1, 1] \end{cases} \quad [4.133]$$

– *Transitional phase and encryption*: in a transitional phase, Trivium is

– iterated 4 × 288 = 1,152 times before generating the first bit of the keystream sequence. This phase allows a sufficient blurring of the keystream. The encryption starts from the 1,115th clock cycle, which corresponds to the first bit generated from the keystream.

– *Performance*:

- *periodicity*: the designers of Trivium indicate after an intuitive analysis of the scheme that the periodicity is at least equal to $2^{93} - 1$. They also show that for a couple of K/IV, the probability of finding a cycle lower than 2^{80} is 2^{-208};

- *bitrate*: in software implementation, using a PC with a CPU at 1.5 GHz, it takes 12 clock cycles to encode one byte, i.e. a bitrate of:

$$D(\text{bit/s}) = \frac{CPU(\text{Hz})}{12} \times 8 = 1 \text{ Gbit/s}$$

For comparison, the standard “AES” encryption algorithm requires 24 clock cycles to encrypt one byte.

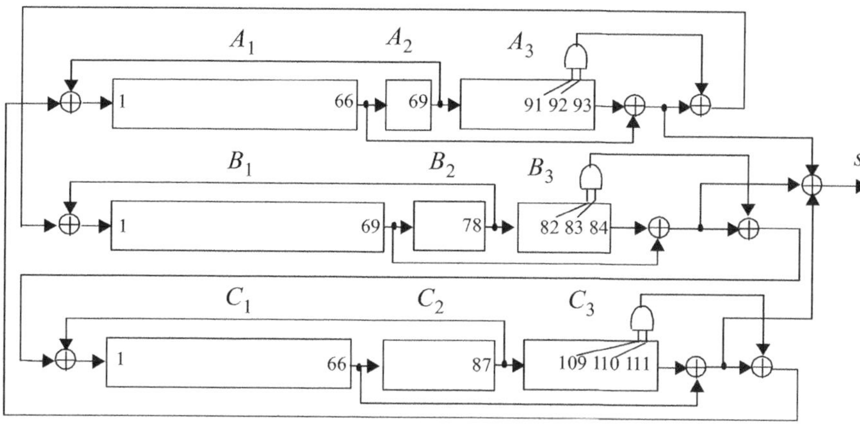

Figure 4.31. *Internal structure of Trivium*

In hardware implementation, using a clock frequency of 125 MHz, Trivium can generate 16 bits per clock cycle, so a bitrate of:

$$D(\text{bit/s}) = CPU(\text{Hz}) \times 16 = 2 \text{ Gbit/s}$$

Since 2007, different types of attacks have been applied to Trivium, thus revealing some weaknesses. However, it provides a good compromise between security and bitrate.

Part 2

Baseband Digital Transmissions and with Carrier Modulation

Introduction to Part 2

We consider here the problem of the transmission of a series of binary and M-ary symbols on a transmission channel characterized by its *limited bandpass of the "low-pass" type and of band-pass type* as well as the associated random disturbances, the *channel noise*.

If the source delivers an analog message such as the speech signal (output of a microphone) or the video signal (output of an analog camera), then it should be digitized. This operation is done by sampling the analog message while respecting Shannon's theorem ($f_e \geq 2f_{max}$), then by quantifying the samples obtained with $Q = 2^m$ levels and finally by coding each sample with m binary elements.

The transmission media of the "low-pass" type are the two-wire and coaxial electrical cables and the optical cables. The bandwidth of twisted pair cables is low. They are generally used for transmissions around 2 Mbit/s or 8 Mbit/s on the telephone network. The bandwidth of coaxial cables is greater than that of two-wire cables. They are used, with optical fibers, to connect telephone exchanges with a bitrate which can reach Gbit/s. Optical fibers are distinguished by a very high bandwidth (several tens of Gbit/s) and a very low attenuation compared to that of metallic cables. They are used for terrestrial networks, for submarine cables (they replace coaxial cables), as well as for distribution networks (connections between telephone exchanges and subscribers).

Channel noise has various origins, such as high-voltage lines, high-speed trains, motors, various radiations, possible interferences between different users of the transmission medium, internal noise from the electronic devices used, etc.

Transmission media of the "bandpass" type are the analog telephone channel, the radio channel, the mobile radio channel and the earth-satellite channel. *Carrier frequency digital transmissions* differ from digital baseband transmissions by the *types* and *properties of the digital modulations* used, as well as by the *types* and *characteristics of radio degradations*.

Finally, note that *digital baseband transmissions* form a *general problem*, the solution of which applies to all forms of digital transmission. Indeed, we will see in Chapter 7 that digital transmissions with carriers, through a bandpass filter, can be reduced to the simpler and more universal case of transmission through an *equivalent low-pass filter*.

Figures I2.1 and I2.2 respectively demonstrate the block diagram of a digital baseband transmission and the general diagram of a digital radio transmission system.

In this second part, we will address the following three issues:

– the binary to M-ary coding and M-ary to signal coding: online codes (Chapter 5);

– the transmission over a low-pass channel of an M-ary digital signal (Chapter 6);

– the digital transmissions with carrier modulation (Chapter 7).

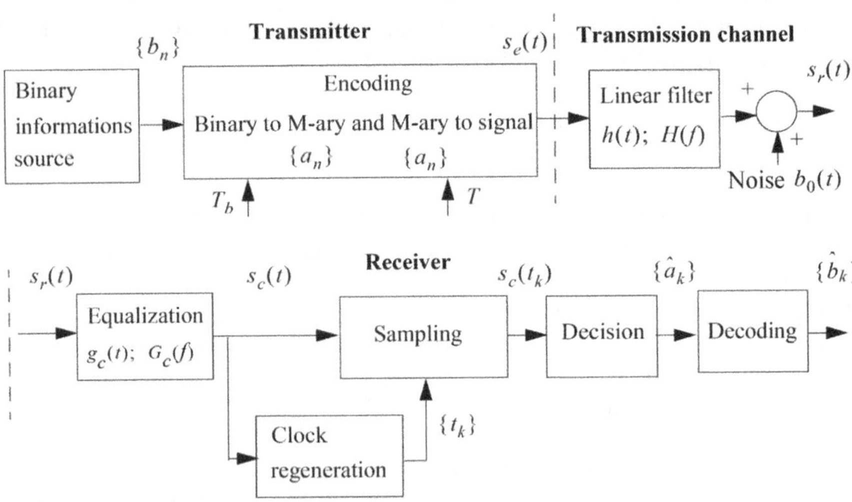

Figure I2.1. *Practical system of a baseband communication*

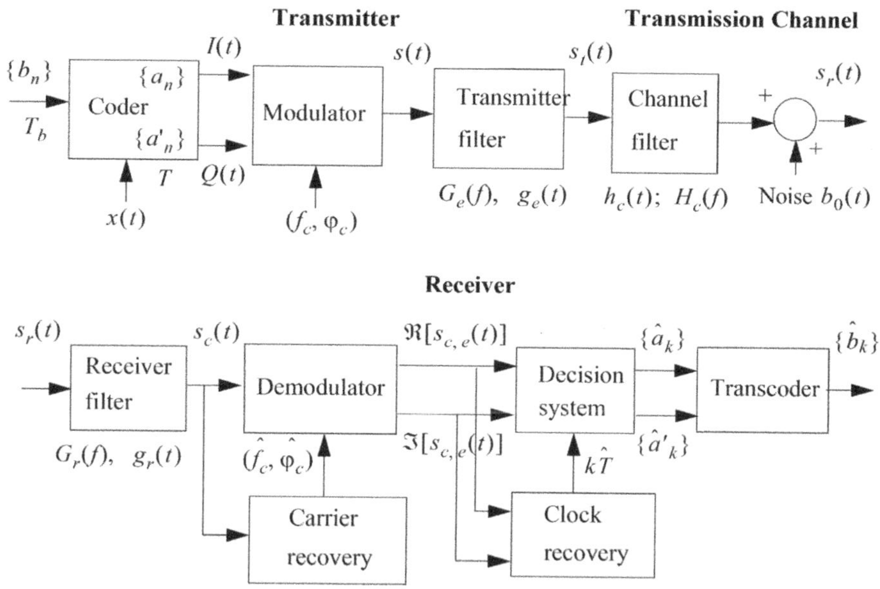

Figure I2.2. *General diagram of a digital radio transmission system*

5

Binary to M-ary Coding and M-ary to Signal Coding: On-line Codes

5.1. Presentation and typology

A digital message, as a sequence of binary or M-ary elements called "information symbols", is an abstract representation. To transmit this message, a physical quantity is associated with it in the form of an electrical signal (or optical signal for transmissions over optical fiber). This chapter describes the on-line codes used for baseband transmissions, as well as their most important properties.

The binary to M-ary transcoding and the M-ary to signal coding consists (see Figure 5.1) in:

– a proper rule which allows one to associate with the binary sequence $\{b_n\}$, $b_n \in [0,1]$ a sequence $\{a_n\}$ where the symbols a_n belong to an alphabet of cardinal $M \geq 2$. This is called transcoding. Its essential purpose is to promote the operation that will follow (on-line code). This rule can be described in several ways: by a dictionary of codewords, by graph or state diagram;

– an on-line coding which consists in constructing from each element a_n a waveform signal which occupies a time slice of at most T (with some exceptions) and which supports the symbol a. The signal thus formed at the output of the encoder is a so-called "baseband" signal. It is written:

$$s_e(t) = \sum_n a_n \, x(t - nT) \qquad [5.1]$$

where $x(t)$ is a deterministic signal of a given pattern on the temporal interval $[0, T[$. A very common pattern for $x(t)$ is that of a rectangular signal (pulse) of V amplitude:

$$x(t) = \begin{cases} V & \forall t \in [0, T[\\ 0 & \text{elsewhere} \end{cases} \quad [5.2]$$

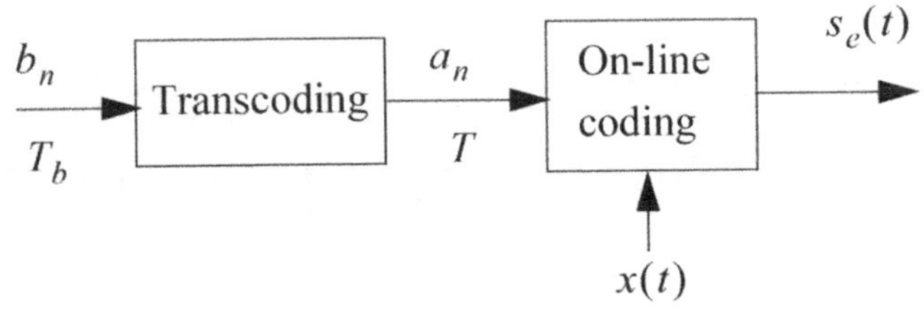

Figure 5.1. *Principle of binary to M-ary transcoding and M-ary to signal coding*

The bitrate D (bit/s) of sequence b_n is defined by the number of bits sent per unit of time. In synchronous digital transmission (our object), the transmission of bits by the source is done at a rate of one bit every T_b seconds:

$$D = 1/T_b \quad [5.3]$$

Usually, binary to M-ary coding consists in associating with each word of k binary elements, a symbol a_n, with:

$$a_n \in \{\pm 1, \pm 3, \cdots, \pm(2p+1), \pm, \cdots, \pm(M-1)\}$$

symmetric M-ary symbols. Knowing that:

$$T = kT_b; \, M = 2^k \quad [5.4]$$

the symbol rate D_s is then related to the bitrate by:

$$D_s = \frac{1}{T} = \frac{1}{kT_b} = \frac{1}{T_b \log_2 M} = \frac{D}{\log_2 M} \quad [5.5]$$

This allows that, at a given bitrate D, the symbol rate is reduced by increasing the number of levels M.

On-line codes can be classified according to the following typology:

– linear codes: NRZ-L, NRZ-M, RZ, polar RZ, biphase (Manchester), biphase-M (differential), and partial response linear codes;

– non-linear codes: non-alphabetic codes (Miller, bipolar RZ (AMI), CMI, HDB-n) and alphabetic codes.

Alphabetic codes will not be described in this book.

5.2. Criteria for choosing an on-line code

For baseband transmissions, the transmission medium consists of a cable (two-wire, coaxial, optical fiber) characterized by a limited bandwidth B of the "low-pass" type. It is necessary to use on-line codes whose characteristics are compatible with the properties of the transmission medium. The choice of a code therefore depends on the good properties that the code must have and which are essentially:

– not to have a continuous component. Thus it will not be necessary to regenerate the continuous component and it will therefore be possible to add a supply voltage to the signal $s_e(t)$ which will make it possible to remotely supply the repeater-regenerators (equipment located at each end of a section of an electric transmission cable);

– to only have a little power at low frequencies because of the significant non-linear distortions and problems related to the separation of the continuous component of remote power supply of the repeater-regenerators with the spectral components of the on-line signal;

– to prevent the spectrum of the transmitted signal from being much wider than the bandwidth B of the transmission channel, despite the pre-equalization which is carried out just before transmission. This also reduces crosstalk;

– to allow the easy recovery of the transmission rhythm (clock recovery) for the synchronization of the receiver on the received signal;

– to allow a transparent transmission: all sequences of binary information are authorized on transmission (for example, for a bipolar code the source cannot emit long sequences of "0" because it would generate at the reception a loss of the rhythm, hence the use of the code HDB-n: high-density bipolar of order n);

– to have a certain redundancy allowing, among other things, one to monitor the quality of the transmission line without knowing the information sequence transmitted (for example, the use of 4B-3T coding).

The criteria for choosing an on-line code depend in part on its spectral properties. It is therefore necessary to know the power spectral density of an on-line code.

5.3. Power spectral densities (PSD) of on-line codes

The power spectral density $\Gamma_{s_e}(f)$ of the signal $s_e(t)$ is calculated in a precise way by the Fourier transform of its autocorrelation function $R_{s_e}(t)$ (generally, this latter is rather difficult to calculate):

$$\Gamma_{s_e}(f) = F\{R_{s_e}(\tau)\} \qquad [5.6]$$

where $R_{s_e}(\tau)$ is given by the following relation:

$$\begin{aligned} R_{s_e}(\tau) &= E[s_e(t) \times s_e(t-\tau)] \\ &= \sum_i \sum_j s_{ei} s_{ej} Pr\{s_e(t) = s_{ei} \text{ and } s_e(t-\tau) = s_{ej}\} \end{aligned} \qquad [5.7]$$

NOTE.– In problems 17 and 18 of the second volume of this book we have detailed, in the general case, the calculation of the autocorrelation function by probabilistic approach and the associated power spectral density of the RZ, NRZ and bipolar RZ on-line codes.

Furthermore, we show that the random data signals used in practice are cyclo-stationary signals (also called periodically "stationary") of period T, so each statistical moment is replaced by its time average calculated over a period T. This will allow us to calculate their power spectral density.

The signal $s_e(t)$ of the relation [5.1] at the output of the encoder can be interpreted as the result of the filtering of a signal $a(t)$ by a filter whose impulse response is $x(t)$:

$$s_e(t) = a(t) \otimes x(t) \qquad [5.8]$$

with:

$$a(t) = \sum_n a_n \, \delta(t - nT) \qquad [5.9]$$

The sign $\otimes$ is the convolution operator.

The power spectral density $\Gamma_{s_e}(f)$ of the signal $s_e(t)$ is then:

$$\Gamma_{s_e}(f) = \Gamma_a(f) \times |X(f)|^2 \qquad [5.10]$$

Where $\Gamma_a(f)$ is the power spectral density of the signal $a(t)$, which will be calculated by the Fourier transform of its autocorrelation function, and $X(f)$ is the Fourier transform of $x(t)$. $\Gamma_a(f)$ makes it possible to have a second degree of freedom (the first degree of freedom being the waveform $x(t)$) to control and shape the power spectral density of the on-line code.

In the case of *partial response codes*, which are part of the linear codes, we show that the calculation of the power spectral density $\Gamma_{s_e}(f)$ of the signal $s_e(t)$ is easily performed from an analogous expression to that given by relation [5.10], as the construction of these codes is carried out from four successive procedures.

Also, we must remember that a code is linear if the coding rule is a digital filtering operation (FIR or IIR filter) or a simple affine transformation.

For example:

$$a_n = \sum_{j=0}^{J-1} h_j \times b_{n-j} \qquad [5.11]$$

The set of coefficients h_j that are integers, $j \in [0, 1, \dots, J-1]$, defines the code.

5.4. Description and spectral characterization of the main linear on-line codes with successive independent symbols

These are codes such as:

$$h_j = \delta_j = \begin{cases} 1 & \text{if } j = 0 \\ 0 & \text{if not} \end{cases} \qquad [5.12]$$

They are more like a technique for shaping pulses than real coding. We also assume for these codes that the binary elements emitted at times nT_b are independent and identically distributed (iid) on the alphabet $\{0, 1\}$ with:

$$p_i = Pr\{b_n = i\};\ i = 0, 1\ \forall\, n \text{ and } p_0 = p_1 = 1/2 \qquad [5.13]$$

The successive symbols a_n are independent because b_n are.

Hereinafter, we describe the main independent successive symbol on-line codes and their spectral characteristics.

NOTE.– On all the chronograms of the various on-line codes that we describe, the bits are taken into account in the middle of the time interval by the clock (Ck) (as in reception) and the power spectral densities are represented for an average power of the on-line code equal to 1 ($P_m = 1$).

5.4.1. *Binary NRZ code (non-return to zero): two-level code, two types of code*

5.4.1.1. *NRZ-L code "level code"*

The NRZ-L code is defined by:

$$\text{if } \begin{cases} b_n = 0 & \rightarrow a_n = -1 \text{ or } 0 \\ b_n = 1 & \rightarrow a_n = 1 \end{cases} \quad [5.14]$$

and:

$$x(t) = \begin{cases} V & \forall t \in [0, T_b[\\ 0 & \text{elsewhere} \end{cases} \quad [5.15]$$

With $a_n = -1$, the emitted signal is symmetrical; if $a_n = 0$, the emitted signal is non-symmetrical.

The diagram of the encoder and decoder of the symmetrical NRZ-L code is given in Figure 5.2.

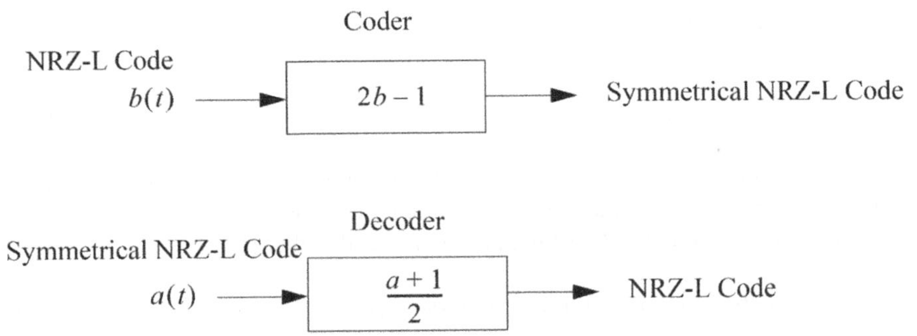

Figure 5.2. *Diagram of encoder and decoder of the symmetrical NRZ-L code*

In Figure 5.3, we present an example of a chronogram of the symmetrical NRZ-L code.

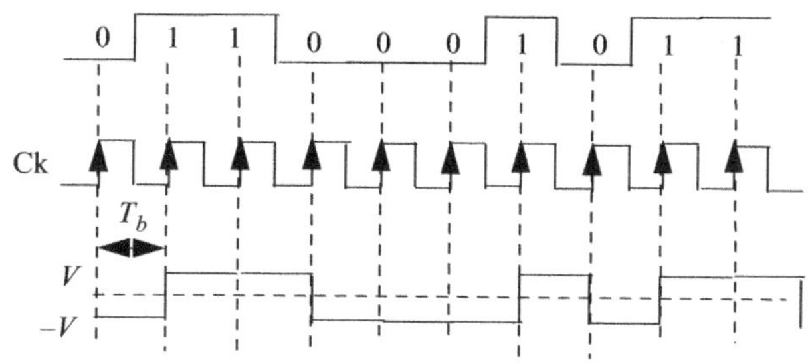

Figure 5.3. *Example of a chronogram of the symmetrical NRZ-L code*

Since the symbols a_n are iid (independent and identical distributed), its mean is zero and its variance σ_a^2 is equal to 1.

The power spectral density of code NRZ-L is given by the following relationships:

$$\Gamma_{NRZ-L}(f) = \frac{V^2 T_b}{4}\left[\frac{\sin(\pi f T_b)}{\pi f T_b}\right]^2 + \frac{V^2}{4}\delta(f) \ \text{ if } a_n = 0 \text{ for } b_n = 0$$

$$\Gamma_{NRZ-LS}(f) = V^2 T_b \left[\frac{\sin(\pi f T_b)}{\pi f T_b}\right]^2 \ \text{ if } a_n = -1 \text{ for } b_n = 0 \qquad [5.16]$$

Figure 5.4 displays the power spectral density of the symmetrical NRZ-L on-line code. It consists of a main lobe which occupies the frequency band $[0,\ 1/T_b]$ (for positive frequencies), and by an infinity of secondary lobes whose amplitude tends towards zero as the frequency increases. In addition, the PSD of this code is canceled at frequencies $f = (k/T)$ with k non-zero integer.

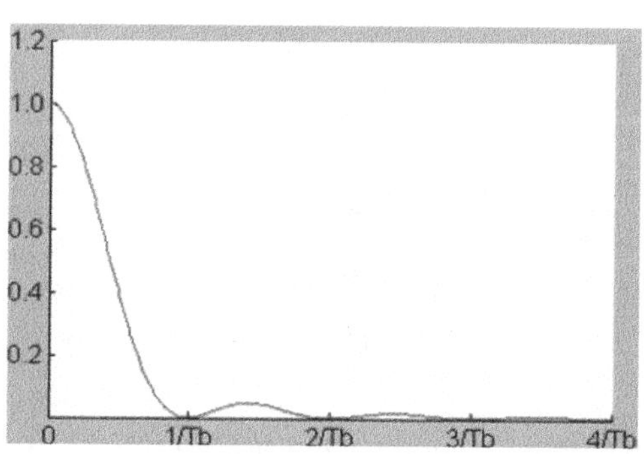

Figure 5.4. *Power spectral density of the symmetrical NRZ-L on-line code. For a color version of this figure, see www.iste.co.uk/assad/digital1.zip*

5.4.1.2. *NRZ-M code "mark" or differential*

The NRZ-M code is defined by the relation [5.17] and Table 5.1:

$$\text{if } \begin{cases} b_n = 0 & \rightarrow a_n = a_{n-1} \\ b_n = 1 & \rightarrow a_n \neq a_{n-1} \end{cases} \quad [5.17]$$

b_n	a_{n-1}	a_n
0	0	0
0	1	1
1	0	1
1	1	0

Table 5.1. *Characterization of the NRZ-M code as a coding and decoding table*

The waveform $x(t)$ and the power spectral density $\Gamma_{NRZ-M}(f)$ are identical to that of the symmetrical NRZ-L code.

In Figure 5.5, we give the block diagram of the on-line NRZ-M coder and decoder and an example of a chronogram of the symmetrical NRZ-M code is shown in Figure 5.6.

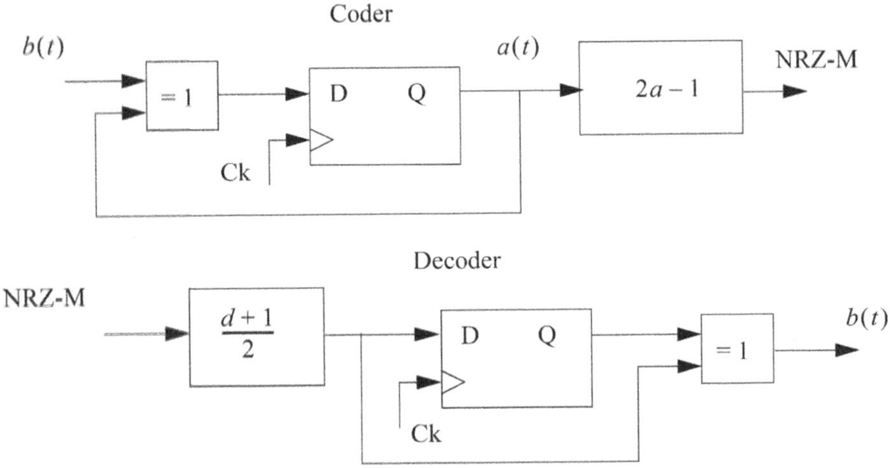

Figure 5.5. *Block diagram of the NRZ-M coder and decoder*

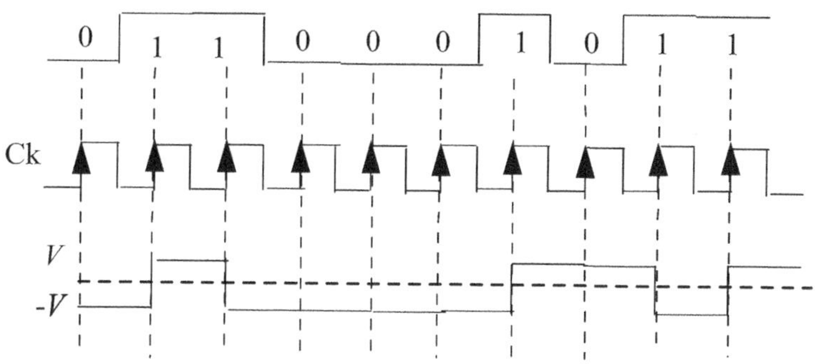

Figure 5.6. *Example of a chronogram of the symmetrical NRZ-M code*

5.4.2. *NRZ M-ary code*

The NRZ M-ary code is a generalization of the binary NRZ code. Each set of k bits from the binary sequence is associated with a symmetrical M-ary symbol:

$$a_n \in \{\pm 1, \pm 3, \cdots, \pm(2p+1), \pm, \cdots, \pm(M-1)\} \text{ with } M = 2^k \qquad [5.18]$$

The waveform $x(t)$ is a "gate function" of amplitude V but of time duration $T = kT_b$.

In the case where the symbols a_n are iid, then their mean is zero and their variance σ_a^2 is equal to:

$$\sigma_a^2 = \frac{2}{M} \sum_{p=0}^{(M/2)-1} (2p+1)^2 = \frac{M^2-1}{3} \qquad [5.19]$$

The power spectral density of the NRZ M-ary on-line code is:

$$\Gamma_{NRZ\ M-ary}(f) = \frac{(M^2-1)}{3} V^2 T \left[\frac{\sin(\pi f T)}{\pi f T}\right]^2 \qquad [5.20]$$

Figure 5.7 shows an example of a chronogram of the NRZ 4-ary code with $T = 2T_b$, $a_n = \{\pm 1, \pm 3\}$ and Figure 5.8 displays its power spectral density $\Gamma_{NRZ\ 4-ary}(f)$.

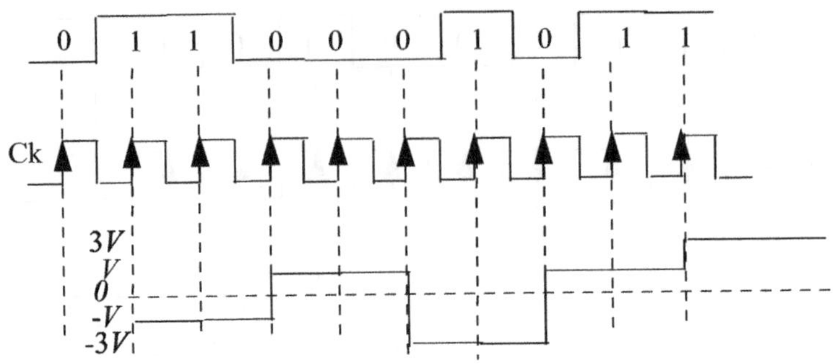

Figure 5.7. *Example of the chronogram of the NRZ 4-ary on-line code*

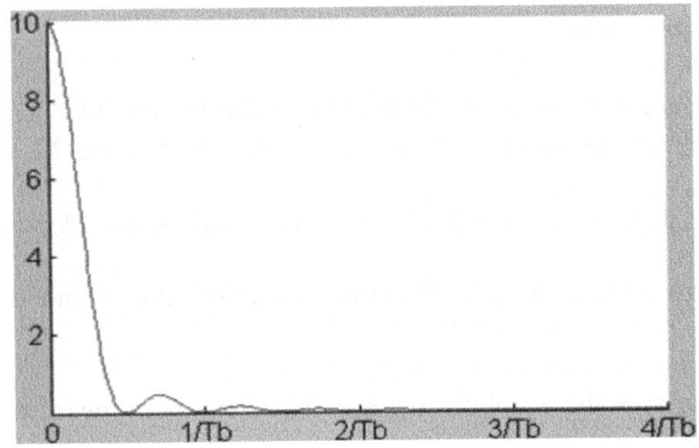

Figure 5.8. *Power spectral density of the symmetrical NRZ 4-ary code. For a color version of this figure, see www.iste.co.uk/assad/digital1.zip*

5.4.3. *Binary RZ code (return to zero)*

This refers to a two-level on-line code such as:

$$\text{if} \begin{cases} b_n = 0 & \to a_n = 0 \\ b_n = 1 & \to a_n = 1 \end{cases} \quad [5.21]$$

and:

$$x(t) = \begin{cases} V & \forall\, t \in [0, \theta T_b[\\ 0 & \forall\, t \in [\theta T_b, T_b[\end{cases} \quad [5.22]$$

with $0 < \theta < 1$, in practice: $\theta = 1/2$.

The binary RZ encoder corresponds to a multiplication operator between the binary sequence and the clock, as shown in Figure 5.9.

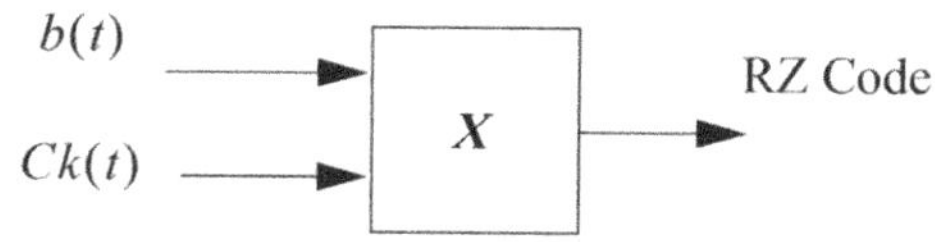

Figure 5.9. *Diagram of the binary RZ on-line code (for θ = 1/2)*

An example of a chronogram (for $\theta = 1/2$) of this code is given in Figure 5.10.

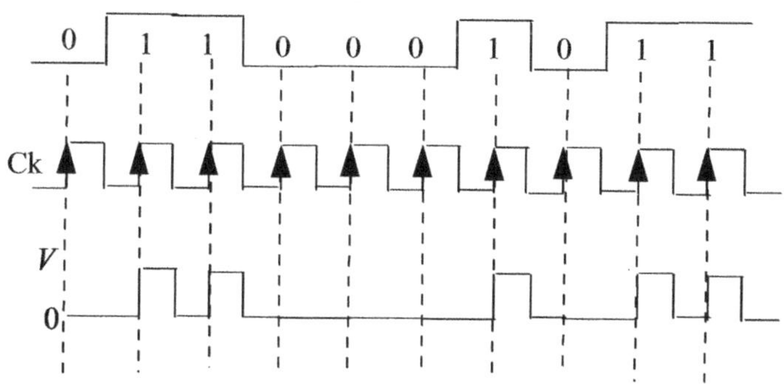

Figure 5.10. *Example of a chronogram of the binary RZ code (for θ = 1/2)*

The symbols a_n being independent and equiprobable, their mean is equal to 1/2 and their variance $\sigma_a^2 = 1/4$. The power spectral density $\Gamma_{RZ}(f)$ is given by the following relation:

$$\Gamma_{RZ}(f) = \frac{V^2\theta^2 T_b}{4}\left[\frac{\sin(\pi f \theta T_b)}{\pi f \theta T_b}\right]^2 + \sum_{k=-\infty}^{\infty} \frac{V^2\theta^2}{4}\left[\frac{\sin(\pi k \theta)}{\pi k \theta}\right]^2 \delta\left(f - \frac{k}{T_b}\right) \quad [5.23]$$

It consists of a continuous spectrum and of a discrete spectrum (discrete frequency components).

In general, one has $\theta = 1/2$, hence:

$$\Gamma_{RZ}(f) = \frac{V^2 T_b}{16}\left[\frac{\sin(\pi f T_b/2)}{\pi f T_b/2}\right]^2 + \frac{V^2}{16}\delta(f) + \frac{V^2}{4\pi^2}\sum_{k=-\infty}^{\infty}\frac{1}{(2k+1)^2}\delta\left(f - \frac{2k+1}{T_b}\right) \quad [5.24]$$

As the relation [5.24] and Figure 5.11 show, the spectral occupancy of this on-line code is double that of the NRZ code and half of the power is located in the discrete components at frequencies $(2k+1)/T_b$. We note in particular a discrete component at the frequency $1/T_b$, which does not have the NRZ code, which allows the recovery of the transmission rhythm (clock) at reception.

For $\theta = 1$, the code is similar to the NRZ code but is asymmetrical, which results in the existence of a discrete component at $f = 0$ (so a non-null continuous component).

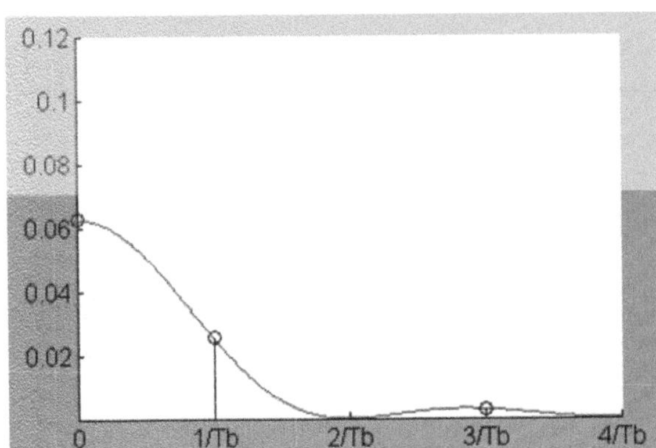

Figure 5.11. *Power spectral density of the binary RZ on-line code (for $\theta = 1/2$). For a color version of this figure, see www.iste.co.uk/assad/digital1.zip*

5.4.4. *Polar RZ on-line code*

This refers to a 3-level on-line code:

$$\text{if}\begin{cases} b_n = 0 & \rightarrow a_n = -1 \\ b_n = 1 & \rightarrow a_n = 1 \end{cases} \quad [5.25]$$

and:

$$x(t) = \begin{cases} V & \forall\, t \in [0, \theta T_b[\\ 0 & \forall\, t \in [\theta T_b, T_b[\end{cases} \qquad [5.26]$$

The block diagram of the encoder is given in Figure 5.12 for $\theta = 1/2$.

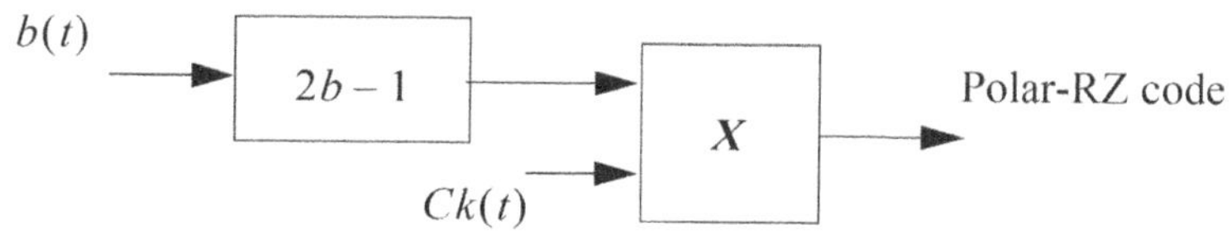

Figure 5.12. *Block diagram of the encoder of the polar RZ on-line code (for θ = 1/2)*

An example of a chronogram for this code is given in Figure 5.13 for $\theta = 1/2$.

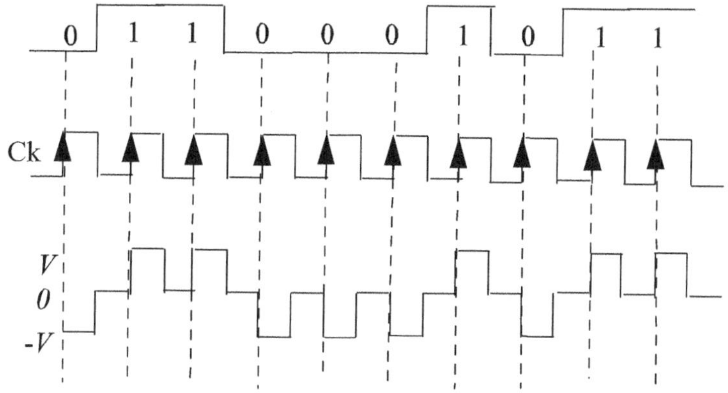

Figure 5.13. *RZ polar code (for θ = 1/2)*

The power spectral density of this code is:

$$\Gamma_{RZ-polar}(f) = V^2\theta^2 T_b \left[\frac{\sin(\pi f \theta T_b)}{\pi f \theta T_b}\right]^2 \qquad [5.27]$$

The spectrum has no discrete components and its continuous part is identical, except to a factor, to that of the RZ codes.

For $\theta = 1/2$, the PSD is given by the relation [5.28]:

$$\Gamma_{RZ-polar}(f) = \frac{V^2 T_b}{4}\left[\frac{\sin(\pi f\, T_b/2)}{\pi f\, T_b/2}\right]^2 \qquad [5.28]$$

It is displayed in Figure 5.14.

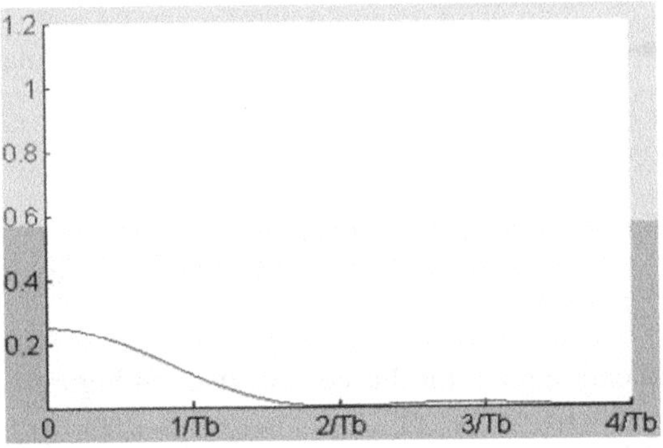

Figure 5.14. *Power spectral density of the polar RZ code (for θ = 1/2). For a color version of this figure, see www.iste.co.uk/assad/digital1.zip*

5.4.5. *Binary biphase on-line code (Manchester code)*

This is a two-level on-line code obtained by transmitting two opposite polarities during the time interval T_b corresponding to a binary symbol, each of these opposite polarities occupying a duration equal to $T_b/2$.

It is defined by:

$$\text{if} \begin{cases} b_n = 0 & \rightarrow a_n = -1 \\ b_n = 1 & \rightarrow a_n = 1 \end{cases} \qquad [5.29]$$

and:

$$x(t) = \begin{cases} V & \forall\, t \in [0, T_b/2[\\ -V & \forall\, t \in [T_b/2\,, T_b[\end{cases} \qquad [5.30]$$

This code always has transitions ($+V, -V$ or vice versa) in the middle of each interval of time duration T_b, regardless of the value transmitted (0 or 1). In addition, for the bit "0" there is a transition at the start of the interval. This is not the case for the bit "1". The transitions are intended to limit the power in the vicinity of frequency zero. They make it possible to recover the clock at reception.

Figure 5.15 displays the block diagram of the binary biphase coder and decoder and Figure 5.16 shows an example of a chronogram of the biphase on-line code.

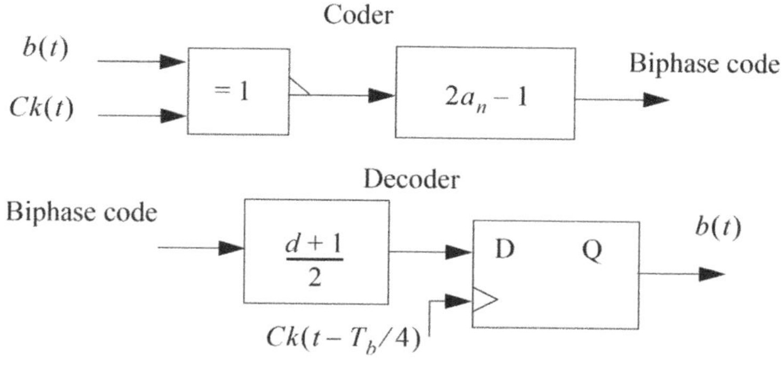

Figure 5.15. *Binary biphase coder and decoder block diagram (Manchester code)*

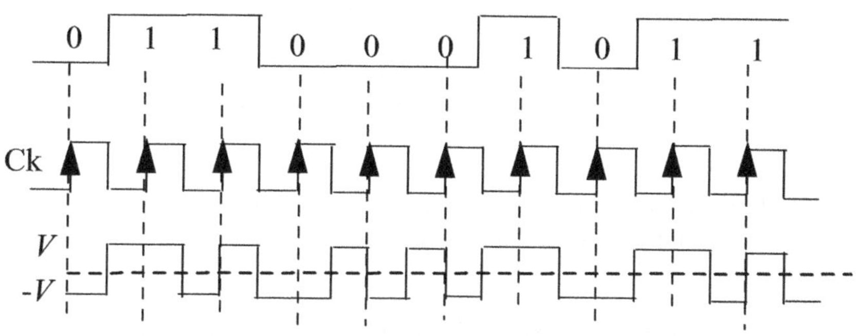

Figure 5.16. *Example of a biphase code chronogram (Manchester code)*

Since the symbols a_n are independent and equiprobable, of zero mean and of unit variance, the power spectral density of the code is:

$$\Gamma_{Biphase}(f) = V^2 T_b \left[\sin\left(\frac{\pi f T_b}{2}\right)\right]^2 \left[\frac{\sin(\pi f\, T_b/2)}{\pi f\, T_b/2}\right]^2 \qquad [5.31]$$

This is displayed in Figure 5.17. It vanishes at zero frequency and remains relatively low for frequencies less than $1/4T_b$. Therefore, this on-line code can be used for relatively long distance cable transmission.

However, the spectral occupancy of the code is double that of the NRZ code.

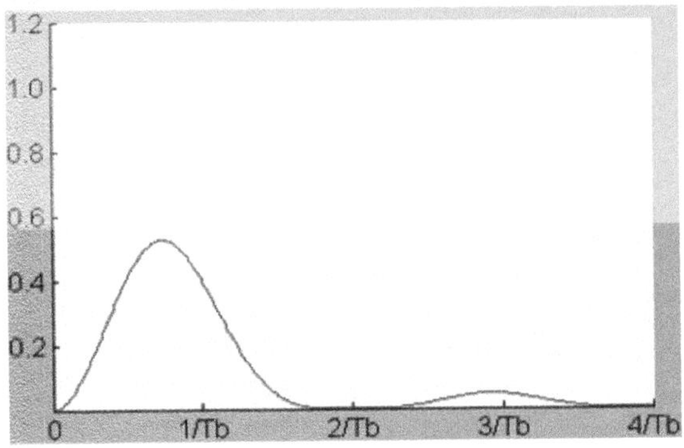

Figure 5.17. *Power spectral density of a biphase code (Manchester code). For a color version of this figure, see www.iste.co.uk/assad/digital1.zip*

5.4.6. *Binary biphase mark or differential code (Manchester mark code)*

The crossing of the transmission wires for a biphase code is equivalent for the receiver to an inversion of polarity and therefore to the inversion of the value of the binary symbols.

To avoid having to locate the line wires, it is preferred to use its differential variant: the differential biphase binary code. In addition, the latter offers better immunity against noise.

Note that all Ethernet systems use this type of on-line code, mainly in the MAC channel access control layer.

The differential biphase code uses the same symbols as the biphase code, but is coded differentially:

$$\text{if } \begin{cases} b_n = 0 & \rightarrow a_n = a_{n-1} \\ b_n = 1 & \rightarrow a_n \neq a_{n-1} \end{cases} \qquad [5.32]$$

and:

$$x(t) = \begin{cases} V & \forall t \in [0, T_b/2[\\ -V & \forall t \in [T_b/2, T_b[\end{cases} \qquad [5.33]$$

The block diagram of the coder and the decoder is given in Figure 5.18 and an example of a chronogram of this code is given in Figure 5.19.

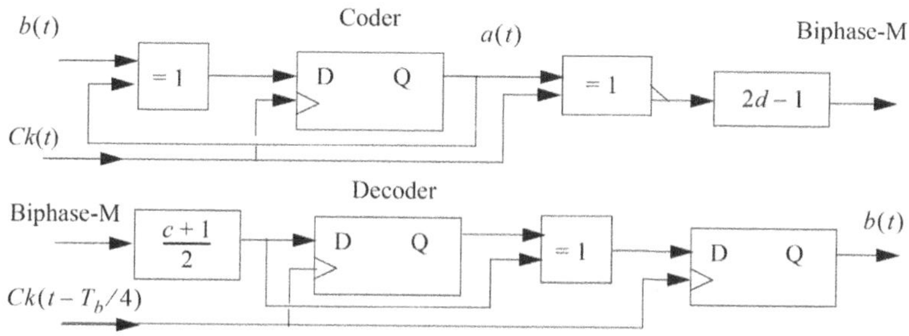

Figure 5.18. *Block diagram of the differential biphase coder and decoder (Manchester mark code)*

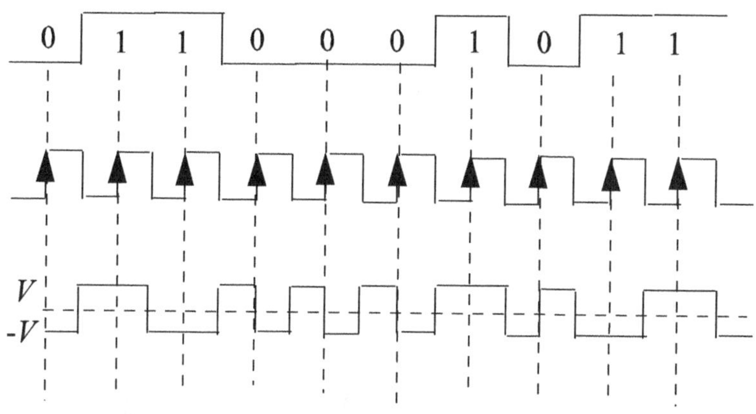

Figure 5.19. *Example of a chronogram of the differential biphase code(Manchester mark code)*

The power spectral density of the code is identical to that of the biphase code, so its spectral occupancy is therefore twice that of the NRZ code.

5.5. Description and spectral characterization of the main on-line non-linear and non-alphabetic codes with successive dependent symbols

These are on-line codes with dependent a_n symbols, although the message source remains with iid binary symbols.

5.5.1. *Miller's code*

The spectral occupancy of this code derived from the biphase code is relatively reduced compared to the previous codes. Miller's code is obtained by "dividing by two" the biphase coded signal. The state change of the Miller code is done on rising edges of the biphase code (from the example of the chronogram of Figure 5.21), but it can also be done on falling edges of the biphase signal.

Decoding consists of detecting the presence or absence of transitions in the middle of the symbol interval. The presence of transition corresponds to a state of the symbol. The absence of transition indicates the opposite symbol state.

The encoder block diagram is given in Figure 5.20 and an example of a chronogram for this code is given in Figure 5.21.

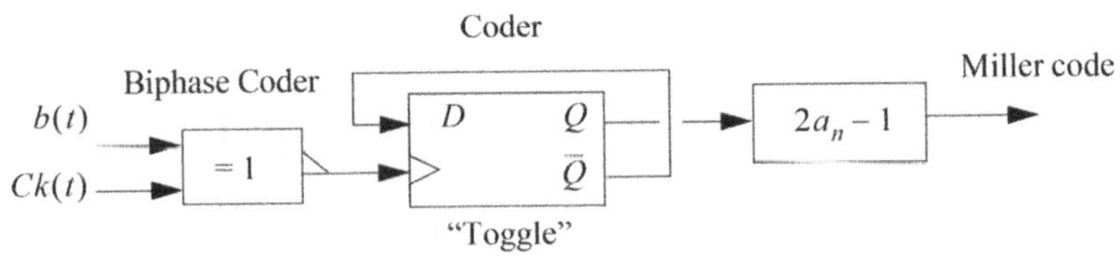

Figure 5.20. *Block diagram of the Miller encoder*

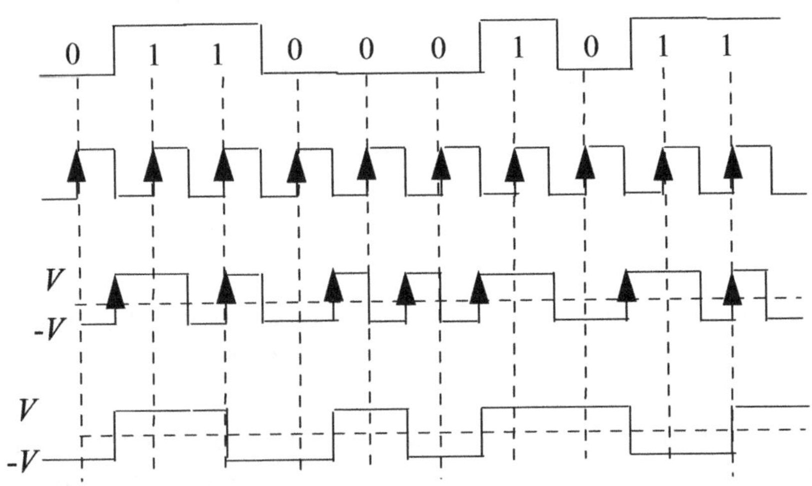

Figure 5.21. *Example of a chronogram of the Miller code*

The power spectral density of this on-line code is:

$$\Gamma_{Miller}(f) = V^2 T_b \left[\frac{1 + 4\cos(\pi f T_b) + 4[\cos(\pi f T_b)]^2 - 4[\cos(\pi f T_b)]^3}{1 - 8[\cos(\pi f T_b)]^2 + 32[\cos(\pi f T_b)]^4}\right] \times \left[\frac{\sin(\pi f\, T_b/2)}{\pi f\, T_b/2}\right]^2 \quad [5.34]$$

It is given in Figure 5.22. It is characterized by a high concentration of power in the vicinity of the frequency $2/5T_b$. The power spectral density (which is here a continuous function) has a low value, but not zero at zero frequency, and the tolerance of the signal to the noise is reduced. However, Miller's code allows the transmission of bitrates up to 5,000 bit/s over telephone cable.

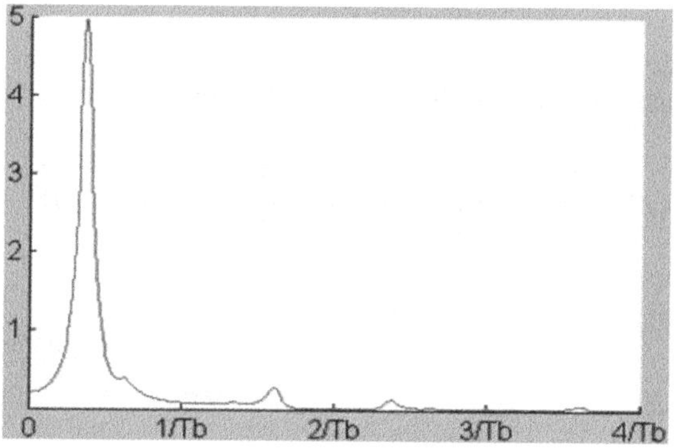

Figure 5.22. *Power spectral density of the Miller on-line code. For a color version of this figure, see www.iste.co.uk/assad/digital1.zip*

5.5.2. *Bipolar RZ code or AMI code (alternate marked inversion)*

This is a three-level on-line code such as:

$$\text{if} \begin{cases} b_n = 0 & \rightarrow a_n = 0 \\ b_n = 1 & \rightarrow a_n = \pm 1 \text{ alternately} \end{cases} \quad [5.35]$$

The correlation of the symbols a_n is carried out by alternately assigning the values +1 and -1 to the symbols a_n when the binary symbol $b_n = 1$.

The mean value of the symbols a_n is zero and their variance σ_a^2 is equal to 1/2.

The signal $x(t)$ is defined by:

$$x(t) = \begin{cases} V & \forall t \in [0, T_b/2[\\ 0 & \text{elsewhere} \end{cases} \quad [5.36]$$

Or by:

$$x(t) = \begin{cases} V & \forall t \in [0, T_b\,[\\ 0 & \text{elsewhere} \end{cases} \quad [5.37]$$

to obtain the NRZ bipolar on-line code.

In Figure 5.23, we show the block diagram of the encoder and decoder of the bipolar RZ code and in Figure 5.24 we present an example of a chronogram of this code.

The power spectral density of the bipolar RZ code is:

$$\Gamma_{Bipolar-RZ}(f) = \frac{V^2}{4} T_b \times \left[\frac{\sin(\pi f\, T_b/2)}{(\pi f\, T_b/2)}\right]^2 \times [\sin(\pi f T_b)]^2 \quad [5.38]$$

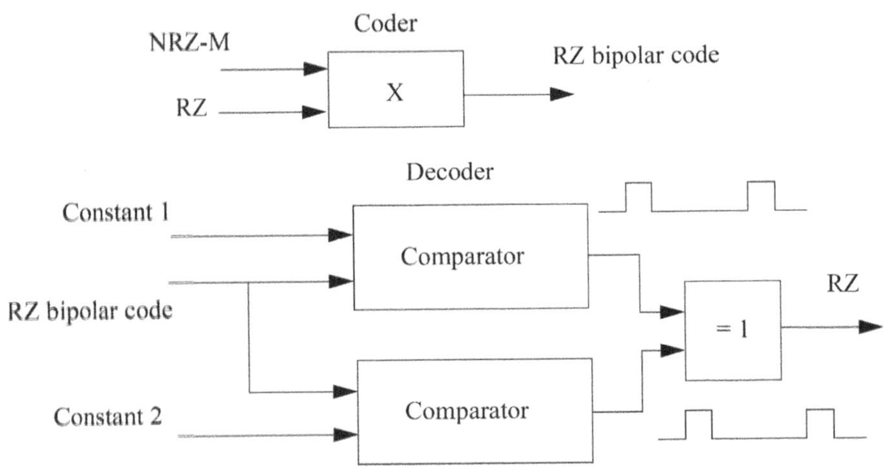

Figure 5.23. *Block diagram of the bipolar RZ encoder and decoder (or AMI)*

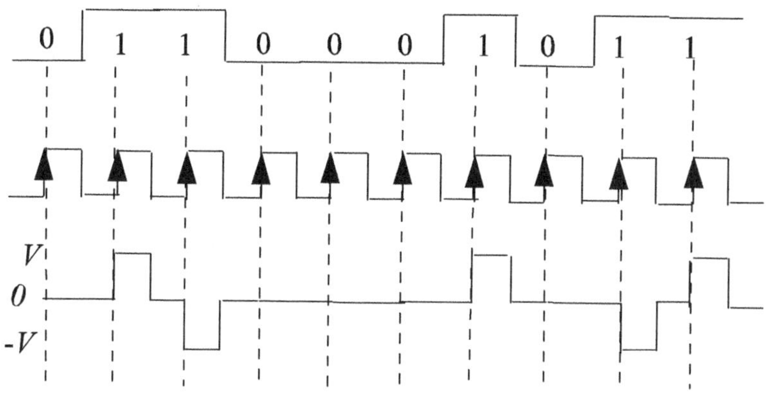

Figure 5.24. *Example of a chronogram of the bipolar RZ code (or AMI)*

Equation [5.38] is shown in Figure 5.25. We notice that the PSD is zero at zero frequency and does not have a discrete component at zero frequency (advantage for remote power supply). The spectral occupancy of the code is practically $1/T_b$.

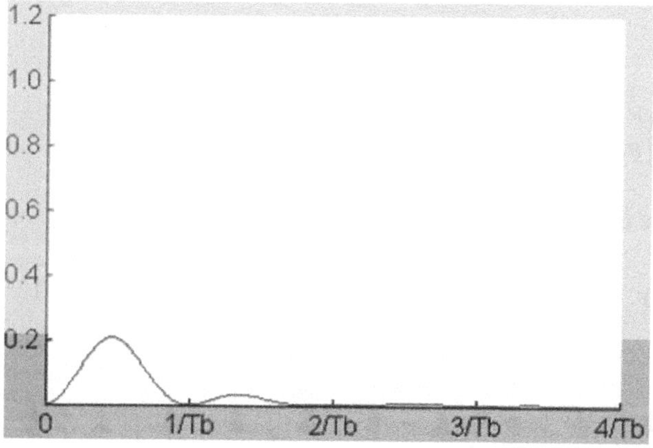

Figure 5.25. *Power spectral density of the bipolar RZ code (or AMI). For a color version of this figure, see www.iste.co.uk/assad/digital1.zip*

Furthermore, by performing a full-wave rectification of the bipolar RZ code, a binary RZ code is obtained which has a discrete component at frequency $1/T_b$ in its power spectral density and, consequently, an easy recovery of the clock at reception.

This code can be used on cables, but since the coding of the binary symbol “0” corresponds to a null signal emitted, this makes it difficult to recover the clock when sending long sequences of “0”, because in this case the amplitude of the frequency component at $1/T_b$ decreases quite strongly and the power of the signal collected after filtering may become insufficient to ensure the recovery of the clock. This has led to the development of on-line codes which have the advantages of the bipolar code, and in addition the advantage of allowing a correct recovery of the clock at reception, whatever the sequence of bits sent by the information source. HDB-n on-line codes are good examples of this type of on-line code (n-order high density bipolar code).

5.5.3. *CMI code (code mark inversion)*

This refers to a two-level code that has the same coding rule as the bipolar code but with different pulse forms:

$$\text{if}\begin{cases} b_n = 0 & \rightarrow \quad a_n = -1, +1 \\ b_n = 1 & \rightarrow \quad a_n = \begin{cases} -1, -1 \\ +1, +1 \end{cases} \text{alternately} \end{cases} \qquad [5.39]$$

and:

$$x(t) = \begin{cases} V & \forall t \in [0, T_b/2[\\ V & \forall t \in [T_b/2, T_b[\end{cases} \qquad [5.40]$$

An example of a CMI code chronogram is given in Figure 5.26.

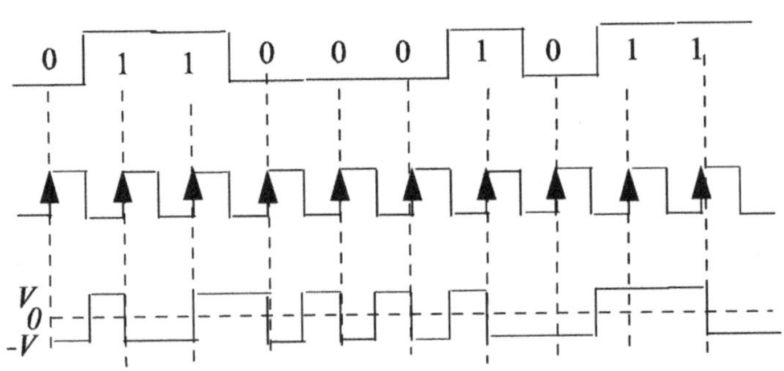

Figure 5.26. *Example of a chronogram of a CMI code*

The power spectral density of this on-line code is:

$$\Gamma_{CMI}(f) = V^2 T_b \left\{ \frac{1}{4} \frac{[\sin(\pi f T_b/2)]^4}{[\pi f T_b/2]^2} + \frac{[\sin(\pi f T_b)]^4}{[\pi f T_b]^2} \right\}$$

$$+ V^2 \sum_{k=-\infty}^{\infty} \frac{1}{\pi^2 (2k+1)^2} \delta\left(f - \frac{2k+1}{T_b}\right) \qquad [5.41]$$

This is shown in Figure 5.27.

It is low at very low frequencies and has a discrete component at the frequency $1/T_b$, so the clock recovery in reception is easy.

The only drawback is its higher spectral occupancy than that of the RZ bipolar code.

The CMI code has been proposed for the interface of 140 Mbit/s digital systems. It is also used in fiber optic transmissions.

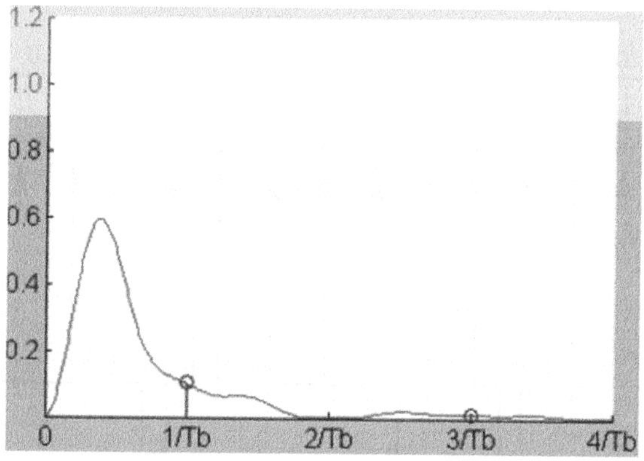

Figure 5.27. *Power spectral density of the CMI on-line code. For a color version of this figure, see www.iste.co.uk/assad/digital1.zip*

5.5.4. *HDB-n code (high density bipolar on-line code of order n)*

It is derived from the bipolar code in which any sequence of $n + 1$ consecutive zeros is replaced by a sequence of at most n consecutive zeros. For this, the $(n + 1)^{\text{th}}$ binary element of the $n + 1$ consecutive zeros sequence is coded by a symbol "V" (violation symbol) equal to ± 1, the sign being chosen in such a way that it violates the rule of alternation. Also, in order to avoid the generation of a non-zero average, the violation symbols must themselves respect the rule of alternation between them.

However, it may be that the receiver cannot recognize a violation symbol as such (if it has a sign contrary to the previous 1); in this case, we code the first 0 of the sequence of $n + 1$ consecutive zeros with a symbol: $a_n = \pm 1$, of the same sign as the violation symbol which follows it. This symbol is called "stuffing symbol B".

IN SUMMARY.– A sequence "$n + 1$" zeros, 0 0..........0 0, is replaced by:

0 0..........0 V or by B 0.........0 V

with:

– V: a symbol that violates the alternation rule of polarity, but satisfies the alternation rule of violation symbols between them;

– B: a stuffing symbol that follows the alternation rule of polarity.

We give in Figure 5.28 an example of a chronogram of the code HDB-3.

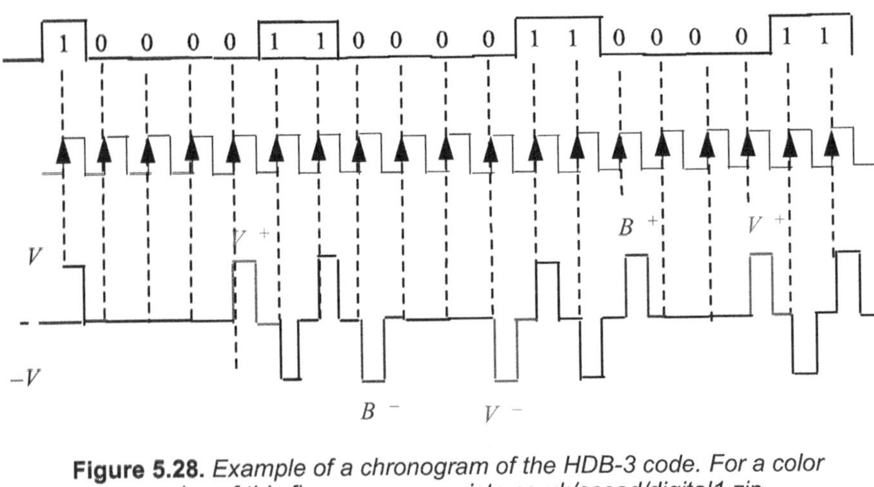

Figure 5.28. *Example of a chronogram of the HDB-3 code. For a color version of this figure, see www.iste.co.uk/assad/digital1.zip*

By doing so, the symbols always have a zero average. The analytical calculation of the power spectral density of the code HDB-n $\Gamma_{HDB-n}(f)$, is complicated. It is comparable to the power spectral density of the bipolar RZ code. Figure 5.29 shows the PSD of the HDB-3 code which is used in the DT1 and DT2 digital transmission systems.

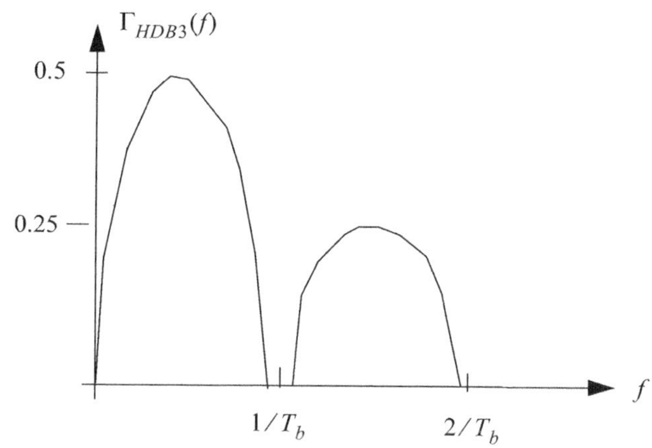

Figure 5.29. *Power spectral density of the HDB-3 on-line code*

In order to easily compare the spectral properties of the different on-line codes studied above, we have represented their power spectral density in Figure 5.30.

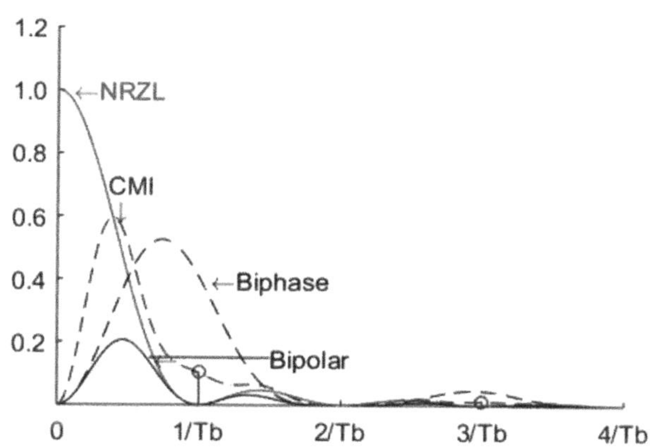

Figure 5.30. *Power spectral density of the main on-line codes presented. For a color version of this figure, see www.iste.co.uk/assad/digital1.zip*

5.6. Description and spectral characterization of partial response linear codes

These are multi-level linear codes that make it possible to easily modify and shape the spectrum of the signal to be transmitted in order to adapt it to the spectral characteristics of a baseband or equivalent low-pass transmission channel.

5.6.1. Generation and interest of precoding

Any partial response linear code can be generated from the sequence b_n by means of four successive operations given in Figure 5.31.

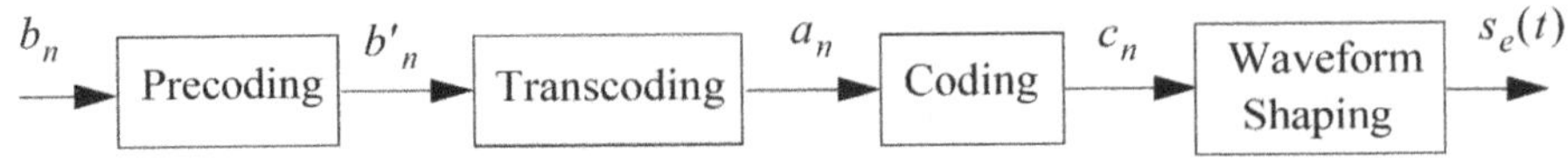

Figure 5.31. *Generation of partial response linear code*

where:

$$\begin{cases} b_n \in \{0,1\} \ ; \ b'_n \in \{0,1\} \ ; \ a_n = 2b'_n - 1, \ a_n \in \{-1,1\} \\ c_n \text{ positive, negative or null integers} \end{cases} \qquad [5.42]$$

NOTE.– For convenience and simplicity, we have limited the alphabet of a_n to two symbols. We could quite easily consider an alphabet with $M = 2^k$ symbols.

– *Precoding*: the pre-coding operation makes the random sequence b_n correspond to a random sequence $b_n^{'}$ of the same bitrate and the same number of levels having certain advantages for decoding. Insofar as the symbols contained in b_n are equiprobable and independent, the same goes for the sequence $b_n^{'}$.

– *Transcoding*: the transcoding operation generates from the random sequence $b_n^{'}$ a random sequence a_n of the same bitrate having a zero mean (symmetrical amplitude of the symbols with respect to zero).

– *Coding*: the partial response linear encoding introduces a controlled inter-symbol dependency, which increases the number of signal levels and changes its spectrum. The encoding is similar to digital filtering. The encoder filter is characterized by its transfer function $H(z)$.

– *Waveform shaping*: waveform shaping is a filtering operation defined by an impulse response $x(t)$ that determines the signal carrying the symbol of the transmitted signal $s_e(t)$:

$$s_e(t) = \sum_n c_n x(t - nT_b) \qquad [5.43]$$

– *Interest of precoding*: the advantage of precoding appears in the decoding procedure. Indeed, if the precoder has a transfer function $P(z)$ which is the reverse of the transfer function $H(z)$ modulo 2 of the coder, we will then have the block diagram of Figure 5.32.

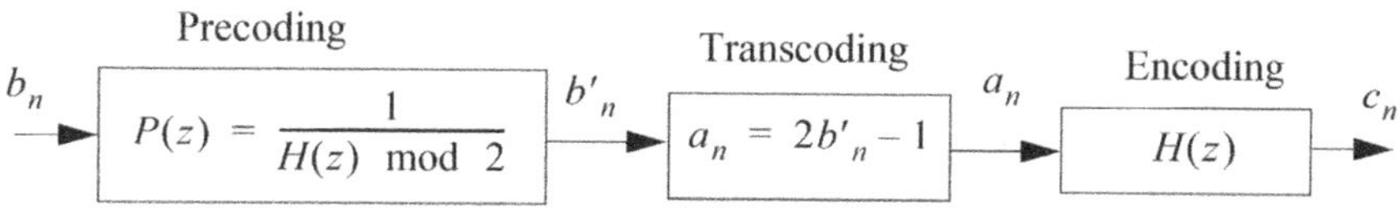

Figure 5.32. *Precoder and encoder pair*

Figure 5.32 shows that one goes from sequence b_n to sequence $b_n^{'}$ by the *Z-transform* filter: $P(z)$. Then conversely, one goes from sequence $b_n^{'}$ to sequence b_n by the *Z-transform* filter: $H(z)$.

So one has:

$$b_n = \sum_{k=0}^{K-1} h_k b'_{n-k} \mod 2 \qquad [5.44]$$

and:

$$c_n = \sum_{k=0}^{K-1} h_k a_{n-k} = \sum_{k=0}^{K-1} h_k [2b'_{n-k} - 1] = 2\sum_{k=0}^{K-1} h_k b'_{n-k} - \sum_{k=0}^{K-1} h_k$$
$$= 2b_n - \sum_{k=0}^{K-1} h_k \qquad [5.45]$$

The coefficients h_k are positive, negative or null integers, and $h_0 = 1$. The FIR (finite impulse response filter) filtering of the encoder corresponds to algebraic operations. The preceding relationships show that on reception and in the absence of transmission noise, we can decode:

$$\hat{b}_n = \frac{1}{2}\left[\hat{c}_n + \sum_{k=0}^{K-1} h_k\right] \mod 2 \qquad [5.46]$$

without needing to use the previous $\hat{b}_{n-k}$.

Otherwise, in the absence of precoding, ($b_n^{'} = b_n$), we would have:

$$c_n = a_n + \sum_{k=1}^{K-1} h_k\, a_{n-k} = 2b_n - 1 + \sum_{k=1}^{K-1} h_k [2b_{n-k} - 1] \qquad [5.47]$$

and decoding would be given by the relation:

$$\hat{b}_n = \frac{1}{2}\left[\hat{c}_n + 1 - \sum_{k=1}^{K-1} h_k[2\hat{b}_{n-k} - 1]\right] \mod 2 \qquad [5.48]$$

Therefore, the estimation of $\hat{b}_n$ would require the estimation of the previous $\hat{b}_{n-k}$. In absolute terms, this would not be a problem, since they have already been estimated before $\hat{b}_n$. However, if they contain errors, then these will propagate.

5.6.2. *Structure of the coder and precoder*

5.6.2.1. *Coder*

The coder is defined by the following relation:

$$c_n = \sum_{k=0}^{K-1} h_k a_{n-k} = a_n + \sum_{k=1}^{K-1} h_k a_{n-k} \qquad [5.49]$$

The *Z-transform* of c_n is:

$$C(z) = A(z) + \sum_{k=1}^{K-1} h_k \times A(z) \times z^{-k} \qquad [5.50]$$

Hence, the transfer function of the filter is:

$$H(z) = \frac{C(z)}{A(z)} = 1 + \sum_{k=1}^{K-1} h_k \times z^{-k} = 1 + \sum_{k=1}^{K-1} h_k \times D^k \qquad [5.51]$$

The coder filter is a FIR filter.

In the literature, the coder is often designated by a polynomial in D (*delay* operator, equivalent to the variable z^{-1} of the digital filtering).

The frequency response of the encoder filter is:

$$H(f) = H(z = \exp[j2\pi f T_b]) = 1 + \sum_{k=1}^{K-1} h_k \exp[-j2\pi k f T_b] \qquad [5.52]$$

$H(f)$ is a periodic function of period $1/T_b$ (T_b is the clock period).

The general structure of the encoder is given by relation [5.49] and presented in Figure 5.33.

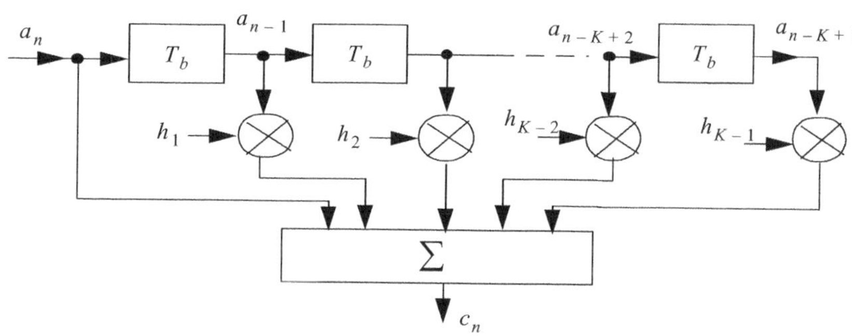

Figure 5.33. *Block diagram of the general structure of the partial response linear encoder*

5.6.2.2. *Precoder*

The precoder is a recursive filter whose transfer function is:

$$P(z) = \frac{1}{H(z)} = \frac{B'(z)}{B(z)} = \frac{1}{1 + \sum_{k=1}^{K-1} h_k z^{-k}} \quad [5.53]$$

One can then write:

$$B'(z) = B(z) - \sum_{k=1}^{K-1} h_k z^{-k} B'(z) \quad [5.54]$$

and:

$$b'_n = b_n - \sum_{k=1}^{K-1} h_k b'_{n-k} \quad \text{mod } 2 \quad [5.55]$$

or:

$$b'_n = b_n \oplus \sum_{k=1}^{K-1} h_k b'_{n-k} \quad \text{mod } 2 \quad [5.56]$$

The structure of the precoder is generally given by relation [5.55], and presented in Figure 5.34. But as the alphabet of b_n and $b_n^{'}$ is $\{0,1\}$, then, in this case, the general structure of the precoder is also given by relation [5.56].

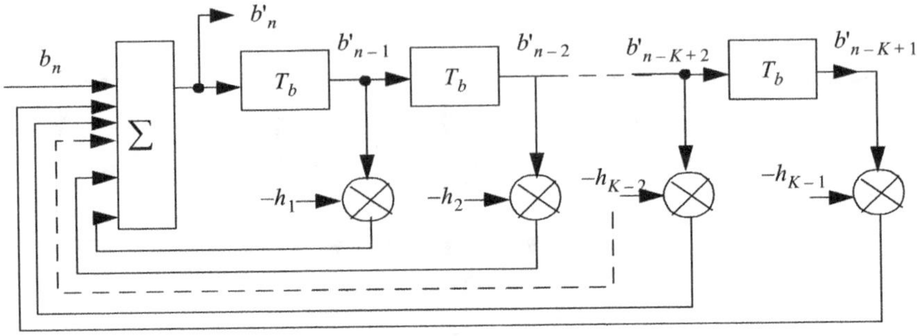

Figure 5.34. *Block diagram of the general structure of the partial response precoder*

5.6.2.3. *Combined precoder and encoder*

By comparing the block diagrams of Figures 5.33 and 5.34, it can be seen that the delay lines of the precoder and encoder can be shared according to the scheme of Figure 5.35.

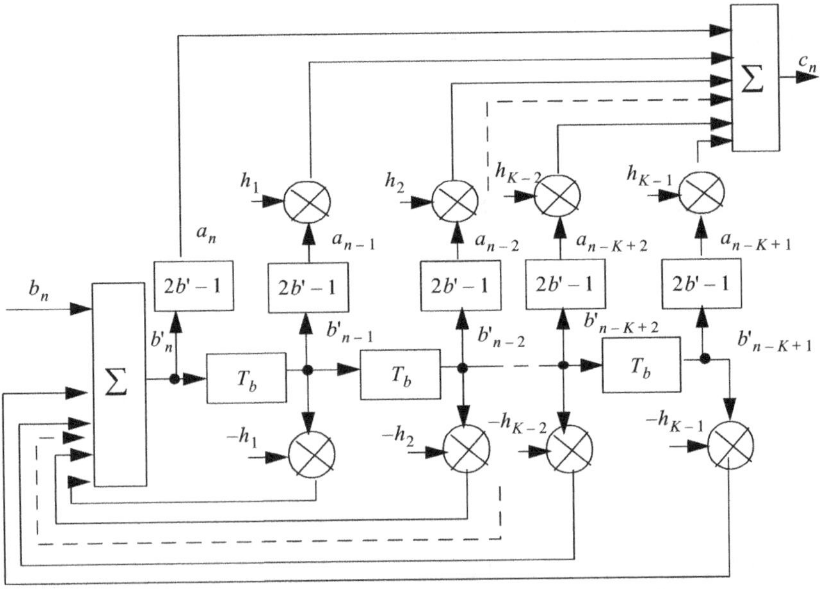

Figure 5.35. *Combined structures of the precoder, transcoder and encoder*

5.6.3. *Power spectral density of partial response linear on-line codes*

To calculate the power spectral density of partial response linear on-line codes, we use the same approach described in section 5.3, by performing this block-by-block calculation according to the scheme of Figure 5.31, but by considering random signals instead of sequences, as shown in the block diagram of Figure 5.36.

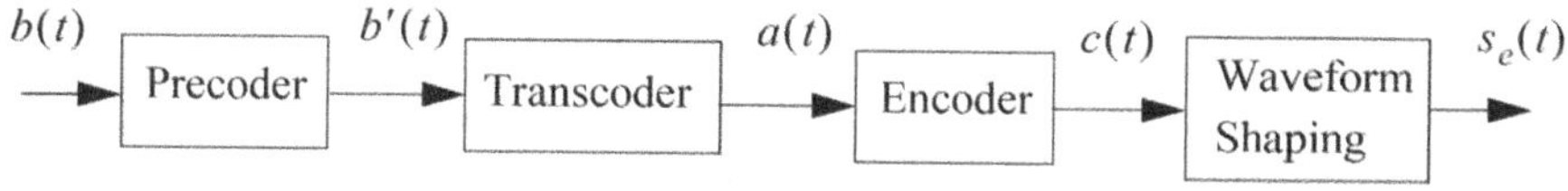

Figure 5.36. *Block diagram of the generation of a partial response linear code for the calculation of the power spectral density*

with:

$$\begin{aligned} b(t) &= \sum_n b_n\,\delta(t-nT_b) \quad b_n \in \{0,1\} \\ b'(t) &= \sum_n b'_n\,\delta(t-nT_b) \quad b'_n \in \{0,1\} \\ a(t) &= \sum_n a_n\,\delta(t-nT_b) \quad a_n = 2b'_n - 1 \quad a_n \in \{-1,1\} \\ c(t) &= \sum_n c_n\delta(t-nT_b) \quad c_n \text{ positive, negative or null integers} \end{aligned} \qquad [5.57]$$

and $\delta(t)$ is the Dirac distribution (sometimes referred to as the "Dirac pulse").

The random signals $b(t)$, $b^{'}(t)$, $a(t)$, $c(t)$, $s_e(t)$ are considered cyclo-stationary and have the power spectral densities: $\Gamma_b(f)$, $\Gamma_b^{'}(f)$, $\Gamma_a(f)$, $\Gamma_c(f)$, $\Gamma_{s_e}(f)$ respectively.

In this case, the autocorrelation function $R_b(\tau)$ of the random sequence of distributions $b(t)$ (b_n are assumed independent and equiprobable: iid) is given by:

$$R_b(\tau) = \frac{1}{4T_b}\delta(\tau) \qquad [5.58]$$

And its power spectral density $\Gamma_b(f)$ is given by:

$$\Gamma_b(f) = F\left\{\frac{1}{4T_b}\delta(\tau)\right\} = \frac{1}{4T_b} \qquad [5.59]$$

The power spectral density $\Gamma_b^{'}(f)$ of the random sequence of distributions $b^{'}(t)$ is identical to that of $b(t)$:

$$\Gamma_b^{'}(f) = \Gamma_b(f) \qquad [5.60]$$

So, the precoding has no influence on the spectral properties of the code.

The autocorrelation function $R_a(\tau)$ of the random sequence of distributions $a(t)$ is given by:

$$R_a(\tau) = \frac{1}{T_b}\delta(\tau) \qquad [5.61]$$

In fact, the general relationship of $R_a(\tau)$ with $a_n \epsilon \{-A, A\}$ is:

$$R_a(\tau) = \frac{A^2}{T_b}\delta(\tau) \qquad [5.62]$$

The power spectral density $\Gamma_a(f)$ of the random sequence of distributions $a(t)$ is:

$$\Gamma_a(f) = \frac{1}{T_b} \qquad [5.63]$$

The power spectral density $\Gamma_c(f)$ of the random sequence of distributions $c(t)$ is:

$$\Gamma_c(f) = \Gamma_a(f) \times |H(f)|^2 = \frac{1}{T_b} \times |H(f)|^2 \qquad [5.64]$$

And finally, the power spectral density of the signal emitted $s_e(t)$ is given by:

$$\Gamma_{s_e}(f) = \Gamma_c(f) \times |X(f)|^2 = \frac{1}{T_b} \times |H(f)|^2 \times |X(f)|^2 \qquad [5.65]$$

This relation shows, on the one hand, that it is possible to shape the spectrum of the signal $s_e(t)$ by modifying $H(f)$ and, on the other hand, that the calculation of the spectrum is easy because it depends on the calculation of $H(f)$ and $X(f)$ separately.

5.6.4. *Most common partial response linear on-line codes*

The most common partial response linear on-line codes, such as duobinary, bipolar NRZ, interlaced bipolar of order 2 and biphase (WAL1 and WAL2), are described hereinafter following the approach used by Michel Stein (Stein 1991). The waveform shaping used is of the NRZ-L type.

5.6.4.1. *Duobinary code*

The duobinary code is characterized by:

$$K = 2\ ;\ h_1 = 1$$

In which case, one has:

$$H(z) = \frac{C(z)}{A(z)} = 1 + z^{-1} = 1 + D \qquad [5.66]$$

hence:

$$c_n = a_n + a_{n-1} \qquad [5.67]$$

The encoder block diagram is given in Figure 5.37.

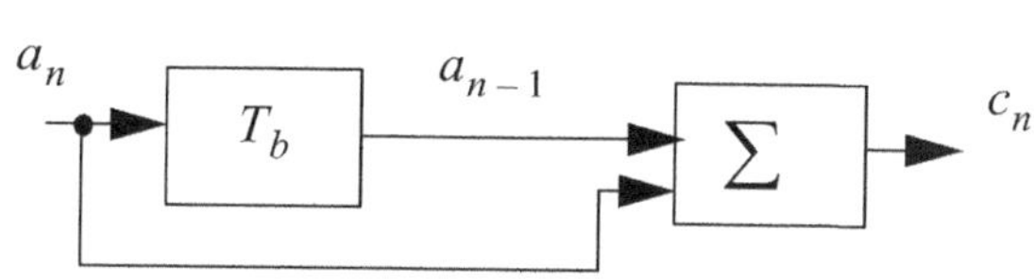

Figure 5.37. *Duobinary encoder block diagram*

The precoder is then defined by the following relation and the block diagram of Figure 5.38:

$$b'_n = b_n \oplus b'_{n-1} \qquad [5.68]$$

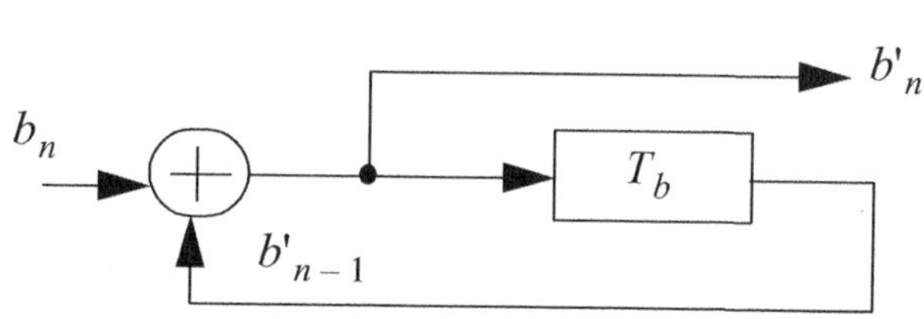

Figure 5.38. *Duobinary precoder*

The combined structure of the precoder, transcoder and duobinary encoder followed by the shaping function is shown in Figure 5.39.

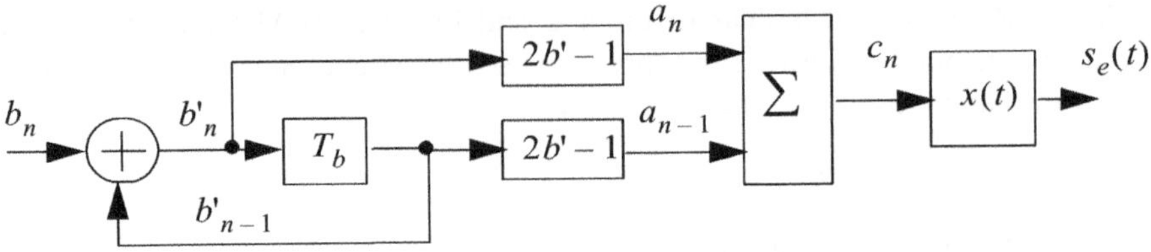

Figure 5.39. *Combined structure of the precoder, transcoder and encoder of the duobinary on-line code*

The graph of an example of a coded sequence is displayed in Figure 5.40.

Note that $\hat{b}_n = \frac{1}{2}\hat{c}_n + 1 \mod 2$ and the sample $s_e(nT)$ of rank n makes it possible to determine $\hat{c}_n$ and then $\hat{b}_n$ (see Chapter 6).

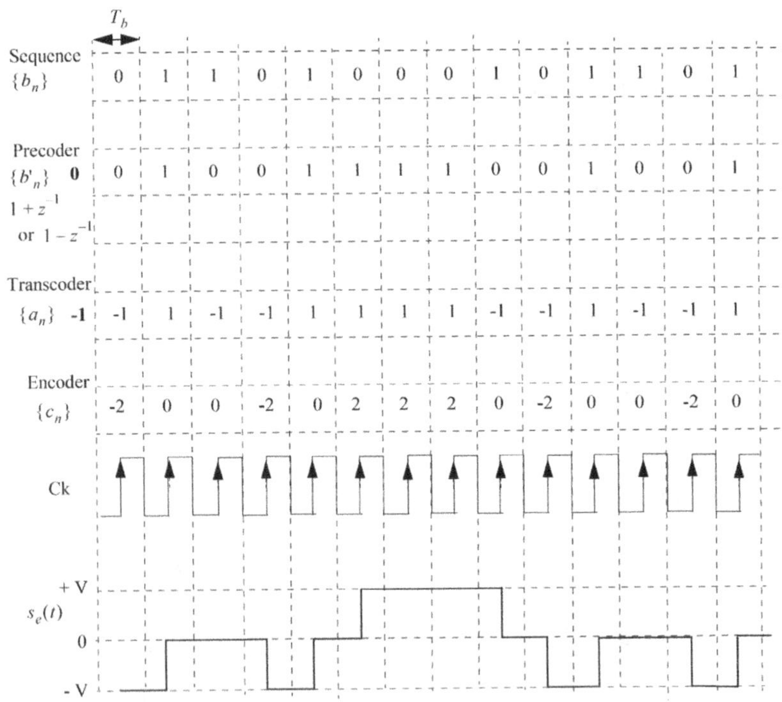

Figure 5.40. *Example of duobinary on-line code*

The spectral properties of the duobinary code are deduced from the transfer function of the encoder:

$$H(z) = 1 + z^{-1} \tag{5.69}$$

hence:

$$H(f) = 1 + \exp(-j2\pi f T_b) = 1 + \cos(2\pi f T_b) - j\sin(2\pi f T_b) \tag{5.70}$$

$$\begin{aligned}|H(f)|^2 &= 1 + 2\cos(2\pi f T_b) + [\cos(2\pi f T_b)]^2 + [\sin(2\pi f T_b)]^2\\ &= 2[1 + \cos(2\pi f T_b)] = 4[\cos(\pi f T_b)]^2\end{aligned} \tag{5.71}$$

If $\pm V$ are the peak values of $s_e(t)$, and the pulse $x(t)$ has an amplitude $V/2$ on $[0, T_b[$, then:

$$X(f) = \frac{V}{2} T_b \times \frac{\sin(\pi f T_b)}{\pi f T_b} \tag{5.72}$$

and from [5.65], the power spectral density is:

$$\begin{aligned}\Gamma_{s_e}(f) &= \frac{1}{T_b} \times |H(f)|^2 \times |X(f)|^2\\ &= \frac{4V^2 T_b}{[2\pi f T_b]^2} \times [\cos(\pi f T_b)]^2 \times [\sin(\pi f T_b)]^2\end{aligned} \tag{5.73}$$

or:

$$\Gamma_{Duobinary}(f) = \Gamma_{s_e}(f) = V^2 T_b \left[\frac{\sin(2\pi f T_b)}{2\pi f T_b}\right]^2 \tag{5.74}$$

It is displayed in Figure 5.41.

The power spectral density has a reduced bandwidth equal to $1/2T_b$ Hz (if we consider the positive frequencies only).

However, with the same peak amplitude, the change from two to three levels increases the sensitivity to noise.

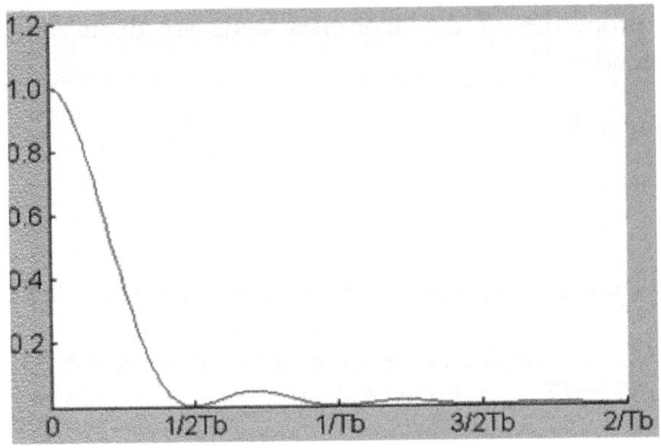

Figure 5.41. *Power spectral density of duobinary on-line code. For a color version of this figure, see www.iste.co.uk/assad/digital1.zip*

5.6.4.2. *NRZ bipolar code*

The bipolar code, the principle of which has been explained above, is seen here as a partial response linear code whose coding law is defined by:

$$K = 2\ ;\ h_1 = -1$$

In which case, one has:

$$H(z) = \frac{C(z)}{A(z)} = 1 - z^{-1} = 1 - D \qquad [5.75]$$

hence:

$$c_n = a_n - a_{n-1} \qquad [5.76]$$

The coder scheme is given in Figure 5.42.

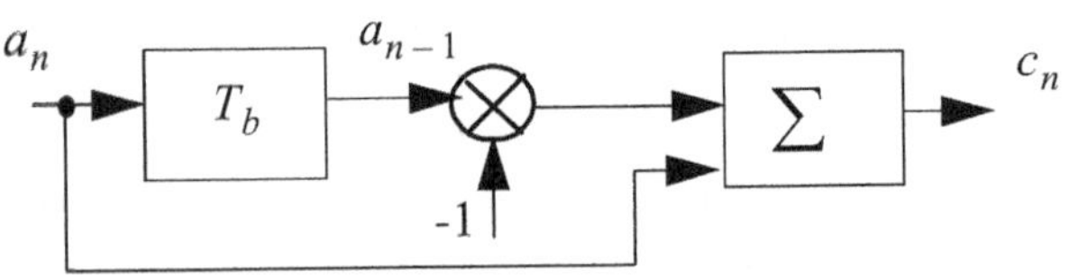

Figure 5.42. *NRZ bipolar code block diagram*

The precoder is identical to that of the duobinary code, and the set of precoder, transcoder, encoder and waveform shaping is shown in Figure 5.43.

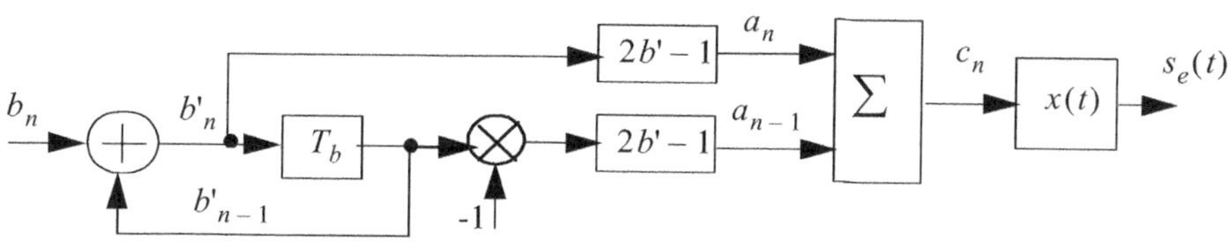

Figure 5.43. *Combined block diagram of the precoder, transcoder, coder and waveform shaping of the NRZ bipolar code*

An example of a coded sequence is shown in Figure 5.44.

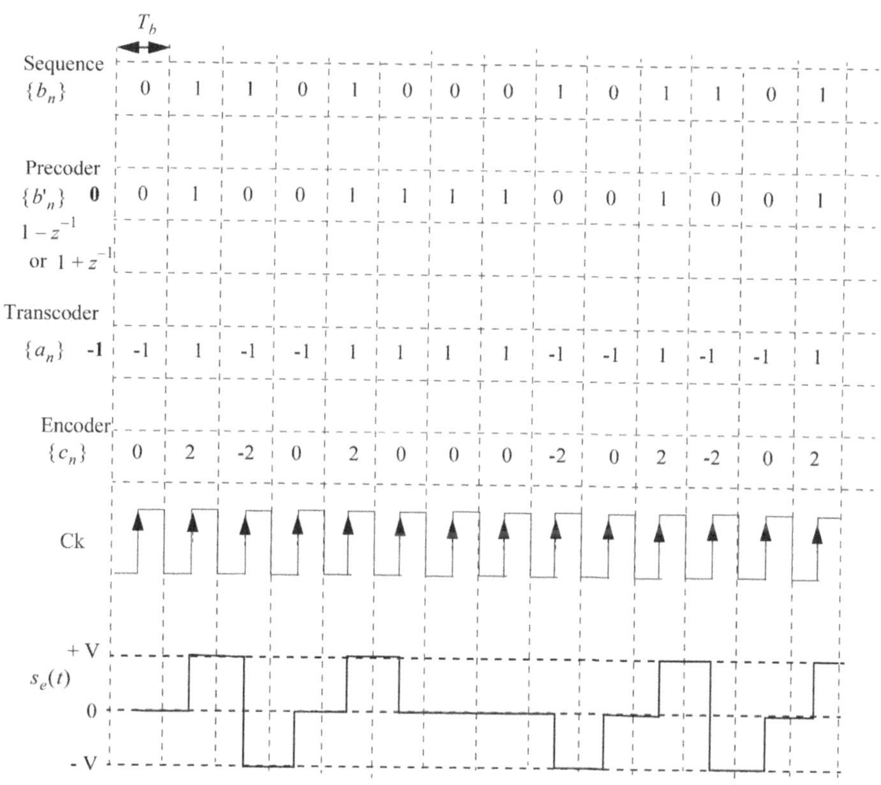

Figure 5.44. *Example of a coded sequence of the NRZ bipolar on-line code*

The transfer function of the encoder is:

$$H(z) = 1 - z^{-1} \quad [5.77]$$

And its power spectral density is calculated as follows:

$$H(f) = 1 - \exp(-j2\pi f T_b) = 1 - \cos(2\pi f T_b) + j\sin(2\pi f T_b) \quad [5.78]$$

$$|H(f)|^2 = 2[1 - \cos(2\pi f T_b)] = 4[\sin(\pi f T_b)]^2 \quad [5.79]$$

hence:

$$\Gamma_{s_e}(f) = \frac{1}{T_b} \times |H(f)|^2 \times |X(f)|^2 = \frac{1}{T_b} \times 4[\sin(\pi f T_b)]^2 \times \frac{V^2 T_b^2}{4}\left[\frac{\sin(\pi f T_b)}{\pi f T_b}\right]^2$$

Finally, one has:

$$\Gamma_{Bipolar-NRZ}(f) = V^2 T_b \times [\sin(\pi f T_b)]^2 \times \left[\frac{\sin(\pi f T_b)}{\pi f T_b}\right]^2 \quad [5.80]$$

This is displayed in Figure 5.45.

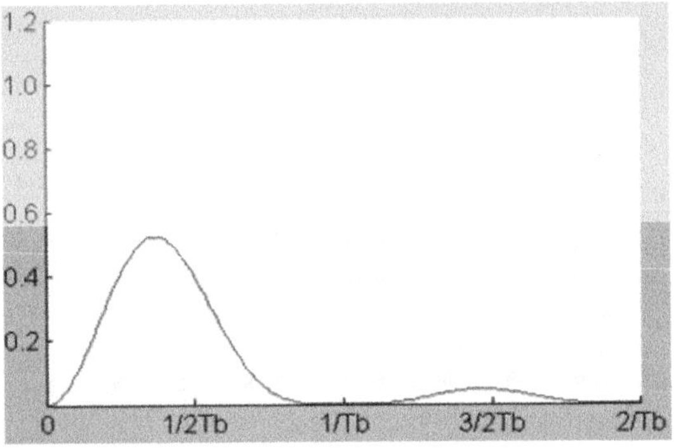

Figure 5.45. *Power spectral density of the NRZ bipolar on-line code. For a color version of this figure, see www.iste.co.uk/assad/digital1.zip*

5.6.4.3. *Interleaved order 2 bipolar code*

The interleaved bipolar coder of order 2 can be obtained by serializing duobinary and bipolar coders. The structure of the 2nd order interleaved bipolar coding chain is then represented in Figure 5.46, where $H_1(z)$ and $H_2(z)$ respectively represent the transfer functions of the duobinary and bipolar coders.

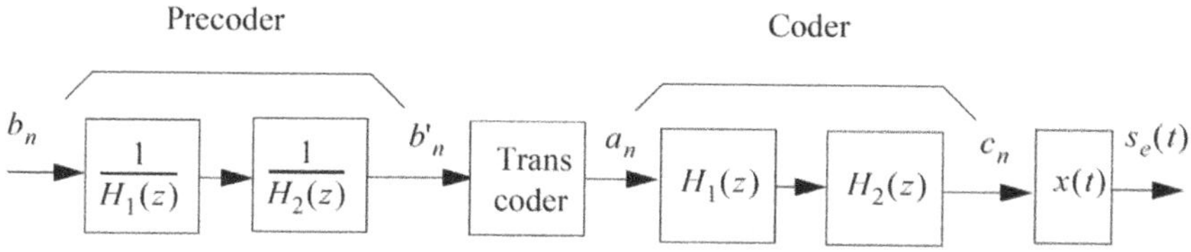

Figure 5.46. *2nd order interleaved bipolar coding structure*

The transfer function and the impulse response of the encoder are written:

$$H(z) = H_1(z) \times H_2(z) \quad [5.81]$$

$$H(z) = (1 + z^{-1}) \times (1 - z^{-1}) = 1 - z^{-2} = 1 - D^2 \quad [5.82]$$

And the output of the encoder is expressed by:

$$c_n = a_n - a_{n-2} \quad [5.83]$$

The encoder block diagram is given in Figure 5.47. It is similar to that of the bipolar coder of Figure 5.42 by replacing the delay T_b by $2T_b$.

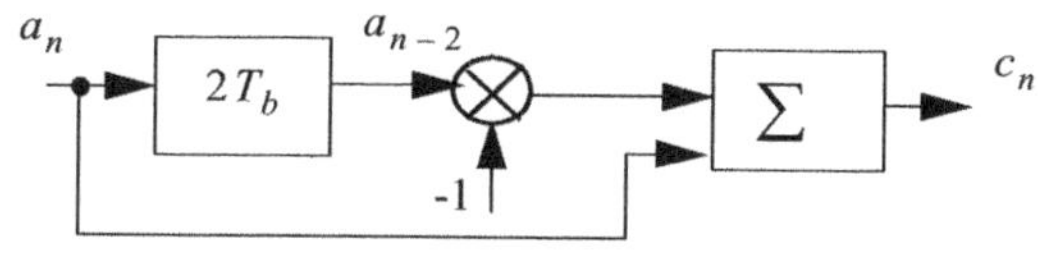

Figure 5.47. *2nd order interleaved bipolar code*

The precoder is defined by the following relation:

$$b'_n = b_n \oplus b'_{n-2} \quad [5.84]$$

Its implementation scheme is shown in Figure 5.48.

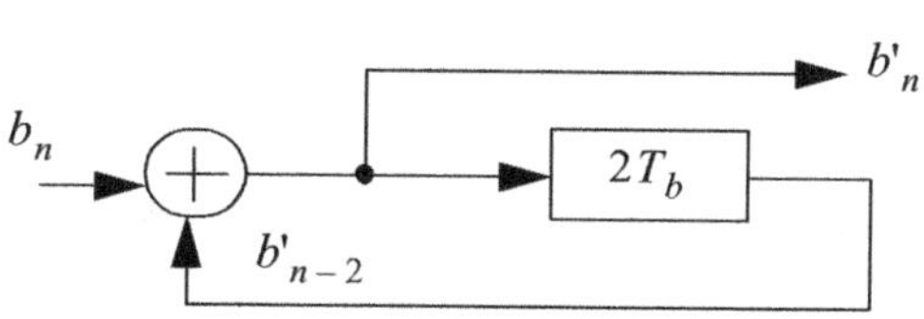

Figure 5.48. *2nd order interleaved bipolar precoder*

The whole chain of two-order interleaved bipolar coding is shown in Figure 5.49.

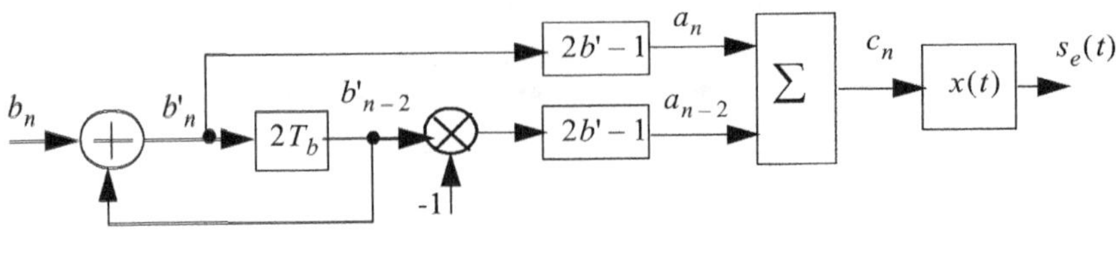

Figure 5.49. *2nd order interleaved bipolar coding chain*

An example of a coded sequence is displayed in Figure 5.50.

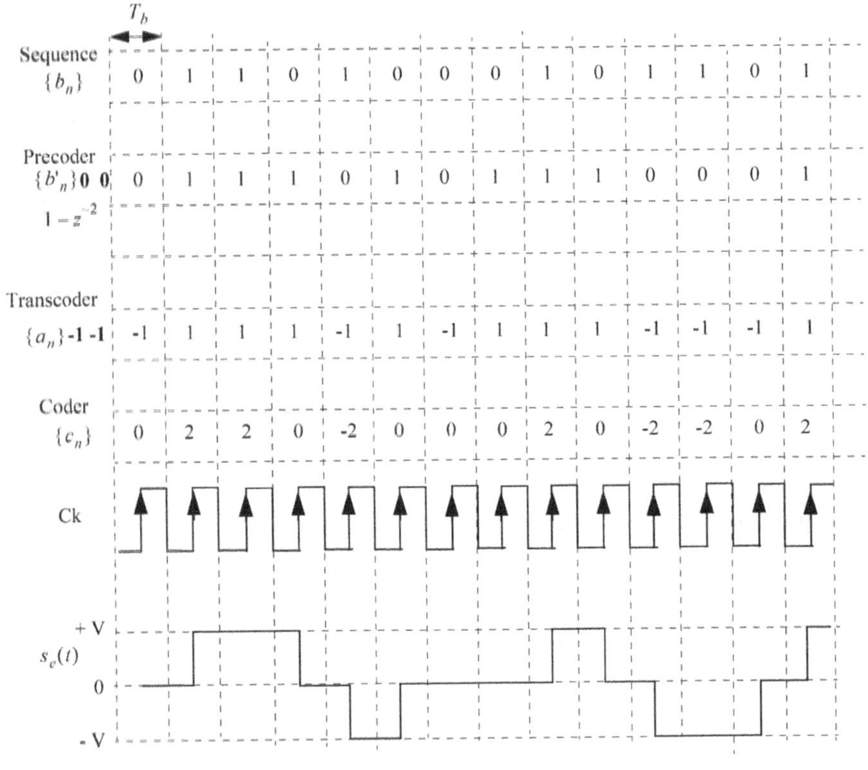

Figure 5.50. *Example of a coded sequence of the 2nd order interleaved bipolar code*

One can easily verify that the coding method illustrated in Figure 5.50 is equivalent to separately carrying out a bipolar coding on the sequence of the even bits $\{b_{2n}\}$ and on the sequence of odd bits $\{b_{2n+1}\}$ of the message to be transmitted $\{b_n\}$. Then, the two coded sequences are interleaved so that the symbol corresponding to the bit of rank n occupies the same rank n in the coded signal.

The frequency response of the encoder is:

$$H(f) = 1 - \exp(-j4\pi f T_b) \quad [5.85]$$

$$|H(f)|^2 = 2[1 - \cos(4\pi f T_b)] = 4[\sin(2\pi f T_b)]^2 \quad [5.86]$$

The power spectral density is then:

$$\Gamma_{s_e}(f) = \frac{1}{T_b}|H(f)|^2|X(f)|^2 = V^2 T_b[\sin(2\pi f T_b)]^2 \times \left[\frac{\sin(\pi f T_b)}{\pi f T_b}\right]^2 \quad [5.87]$$

This is displayed in Figure 5.51.

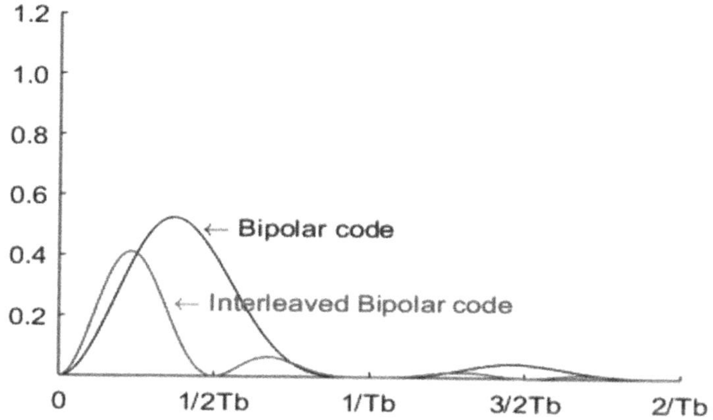

Figure 5.51. *Power spectral density of the 2nd order interleaved bipolar codes and the simple bipolar code. For a color version of this figure, see www.iste.co.uk/assad/digital1.zip*

The spectral occupancy of the interleaved bipolar code of order 2 is half that of the simple bipolar code.

5.6.4.4. *Biphase codes*

These codes are characterized by the inversion or not of a periodic carrier signal as a function of the binary element to be transmitted. The most used biphase code is that described previously and designated here by the term "WAL1". Another variant, called "WAL2" can be used in which the carrier signal has a phase shifted by $\pi/2$ with respect to the series of bits. Figure 5.52 shows the two basic possible forms of signals in each variant.

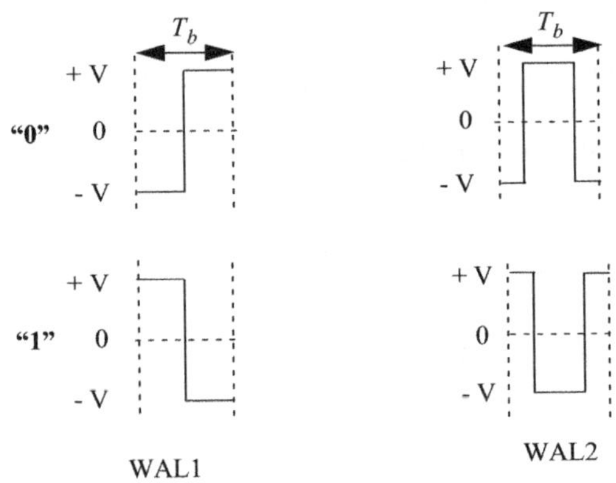

Figure 5.52. *Basic signals used in biphase codes WAL1 and WAL2*

The precoding operation for these codes is replaced by the identity transformation, therefore:

$$\{b'_n\} = \{b_n\} \quad [5.88]$$

5.6.4.5. *Biphase code WAL1*

Its transfer function is given by:

$$H(z) = -1 + z^{-1/2} \quad [5.89]$$

and $H(f)$ is given by:

$$H(f) = -1 + \exp(-j\pi f T_b) \quad [5.90]$$

then:

$$|H(f)|^2 = 2[1 - \cos(\pi f T_b)] = 4[\sin(\pi f\, T_b/2)]^2 \quad [5.91]$$

The shaping filter has a rectangular impulse response of width $T_b/2$. Its transfer function is therefore:

$$X(f) = \frac{VT_b}{2} \times \frac{\sin(\pi f\, T_b/2)}{\pi f\, T_b/2} \quad [5.92]$$

Hence, its power spectral density $\Gamma_{s_e}(f)$ is given by:

$$\Gamma_{s_e}(f) = \frac{1}{T_b}|H(f)|^2|X(f)|^2 = V^2T_b[\sin(\pi f\, T_b/2)]^2 \times \left[\frac{\sin(\pi f\, T_b/2)}{\pi f\, T_b/2}\right]^2 \quad [5.93]$$

The differential biphase code described above is obtained by putting a precoder defined by the following equation upstream of the coder:

$$b'_n = b_n \oplus b'_{n-1} \quad [5.94]$$

5.6.4.6. *Biphase code WAL2*

The coding and shaping functions are:

$$H(z) = -1 + z^{-1/4} + z^{-1/2} - z^{-3/4} = \left(-1 + z^{-1/2}\right) \times \left(1 - z^{-1/4}\right) \quad [5.95]$$

$$H(f) = [-1 + \exp(-j\pi f T_b)] \times [1 - \exp(-j\pi f\, T_b/2)] \quad [5.96]$$

$$|H(f)|^2 = 16[\sin(\pi f\, T_b/2)]^2 \times [\sin(\pi f\, T_b/4)]^2 \quad [5.97]$$

and:

$$X(f) = \frac{VT_b}{4} \times \frac{\sin(\pi f\, T_b/4)}{\pi f\, T_b/4} \quad [5.98]$$

So one has:

$$\Gamma_{s_e}(f) = \frac{1}{T_b}|H(f)|^2|X(f)|^2 = 4V^2T_b[\sin(\pi f\, T_b/4)]^4 \times \left[\frac{\sin(\pi f\, T_b/2)}{\pi f\, T_b/2}\right]^2 \quad [5.99]$$

The same precoder as for the WAL1 code allows differential coding.

The power spectral densities of the two-phase codes WAL1 and WAL2 are displayed in Figure 5.53.

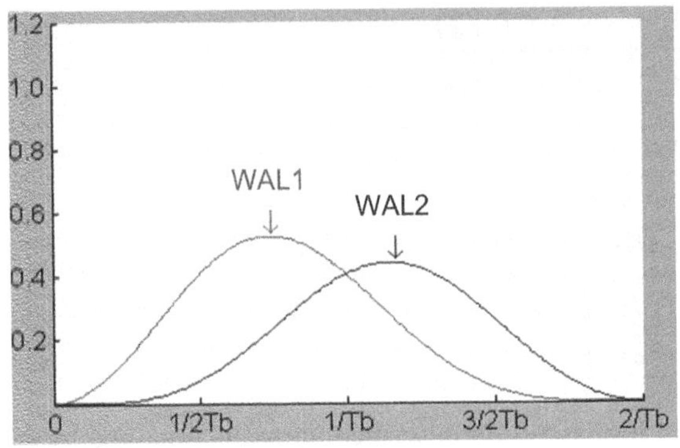

Figure 5.53. *Power spectral densities of the two-phase codes WAL1 and WAL2. For a color version of this figure, see www.iste.co.uk/assad/digital1.zip*

We note that the PSD of code WAL2 has very little energy at low frequencies and this part of the spectrum remains available for useful purposes (adding a voice channel, for example).

Finally, we have also displayed, in Figure 5.54, the power spectral density of the different partial response linear codes analyzed above, in order to allow an easy comparison of the power spectral density properties of these codes.

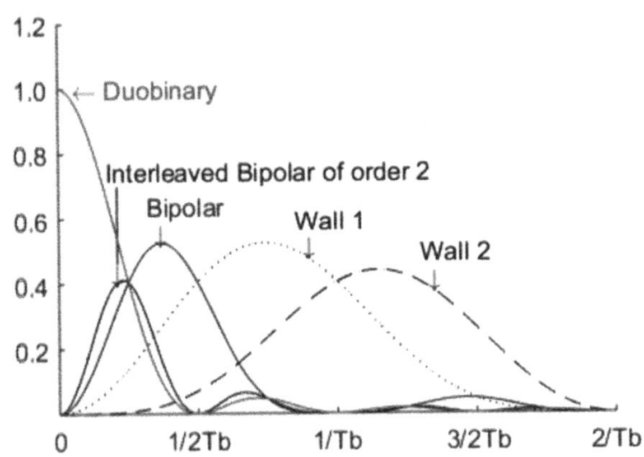

Figure 5.54. *Power spectral density of on-line codes presented. For a color version of this figure, see www.iste.co.uk/assad/digital1.zip*

6

Transmission of an M-ary Digital Signal on a Low-pass Channel

6.1. Introduction

To transmit a digital signal without distortion on a channel (only attenuation α and a given delay τ are accepted), the bandwidth of the transmission medium must be very large compared to the spectrum of the transmitted digital signal. The latter is, in any case, limited to a frequency band Δf by filtering the signal at the output of the on-line encoder.

In practice, the bandwidth of the channel is limited. This limitation can come from the transmission medium itself, which can be frequency selective (all the spectral components of the modulated signal do not propagate at the same phase velocity, because of variable attenuation as a function of frequency). This is precisely the case for two-wire and coaxial metal cables, which have a frequency dependent attenuation $\cong \sqrt{f}$.

The determining parameters of electric cables with regard to signal transmission are, for a given reference length:

– frequency transfer function (attenuation and phase shift);

– characteristic impedance;

– group propagation delay.

Before tackling the problem of digital transmission, in the next section we present digital systems and standardization for high data rate transmissions.

6.2. Digital systems and standardization for high data rate transmissions

In digital transmission systems, the transmission is actually simultaneous on several information channels. They are characterized by different important features such as:

– the *bitrate in bit/s*, which measures the number of binary elements transmitted per unit of time (usually per second);

– the *symbol rate in symbol/s* (or modulation speed, in bauds), which measures the number of M-ary symbols transmitted per unit of time;

– the *number of information channels* transmitted simultaneously;

– the *structure of the frame*, which characterizes the sequential organization of the different channels as well as the auxiliary channels necessary for *signaling* and *locking the frames* (frame synchronization);

– the *transmission parameters*: the type of transmission medium, the average length of a section, the probability of errors, etc.

For practical reasons, as well as for theoretical and economic reasons, the bitrates found in the digital transmission systems used are not arbitrary but correspond to a digital hierarchy depending on standardization. The digital transmission network, using copper cables as a medium, is based on the use of the so-called "plesiochronous" digital hierarchy (PDH). The word "plesiochronous" comes from the Greek *plesio* (near) and *chronos* (time) and reflects the fact that PDH networks use identical but not perfectly synchronized elements: they have the same nominal bitrate for all arteries of the same type, but this bitrate differs slightly depending on the local processing clock. In this hierarchy, there are five types of digital transmission systems: DT1, DT2, DT3, DT4 and DT5, where DTi means "digital transmissions of order i". It should be noted that a DTi system can transport four DT (i–1) systems simultaneously. The transition from one level to another and vice versa is done by multiplexing and demultiplexing. In addition, transmission systems can be found which transport several hierarchies of the same order simultaneously (for example, $n \times 34$ Mbits or $n \times 140$ Mbits). Note that the bitrates are different in the USA and Japan.

Table 6.1 gives the main features of the five digital transmission systems standardized in Europe.

Designation	Bitrate Mbit/s	Related multiplex	Number of telephone channels	Code and frame length	Medium	Nominal length of regeneration
DT1	2.048 (2)	30 x 64Kbit/s + 128 Kbit/s	30/60 channels or 5 channels Conference call or 15 channels Video-conference	HDB3 256 bits (240 useful)	Cable of symmetric pairs ordinaries	1.8 km
DT2	8.448 (8)	4 x DT1	120/240 channels or 20 channels Conference call or 60 channels Video-conference	AMI 848 bits = 4 x 212 (824 useful)	Cable of symmetric pairs specials or mini coaxial 0.7/2.9 mm	4 km ± 400 m
DT3	34.368 (34)	4 x DT2	480/960 channels or 4 TV channels	–	Cable of coaxial pairs 1.2/4.4 mm	4 km ± 100 m
DT4	139.264 (140)	4 x DT3	1,920/3,840 channels or 16 TV channels	4B3T	Cable of coaxial pairs 1.2/4.4 mm or 2.6/9.5 mm	2 km ± 50 m
DT5	564.992 (565)	4 x DT4	7,680/15,360 channels or 64 TV channels	–	Cable of coaxial pairs 2.6/9.5 mm	1.5 km

Table 6.1. *Main features of the five digital transmission systems standardized in Europe*

In addition, the flexibility requirements of the transmission network, the necessary improvement in operating and maintenance functions, the appearance of transmissions on optical fibers and the rapid increase in their transmission capacity, the necessary interconnections between networks to high and standardized data rates from different operators have led to the study and standardization of a new synchronous digital hierarchy: SDH (Synchronous Digital Hierarchy).

The use of PDH is most often limited to 140 Mbit/s, after which SDH is preferred.

6.3. Modeling the transmission of an M-ary digital signal through the communication chain

Let us consider the transmission of an M-ary digital signal through a noisy low-pass channel of bandwidth B (see Figure 6.1).

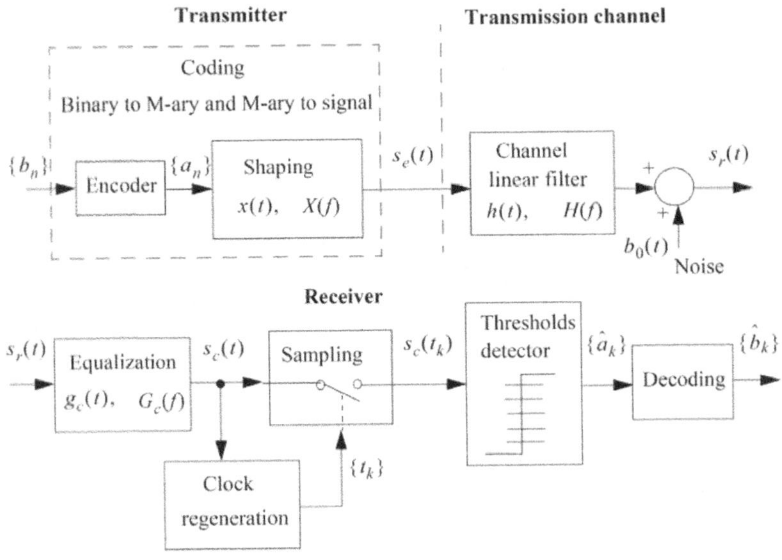

Figure 6.1. *Practical chain of a digital baseband communication system*

The transmission channel is modeled by a low-pass filter having the following known attenuation, phase shift and delay characteristics:

$$\begin{cases} H(f) = |H(f)|\exp[j\phi_h(f)] \quad \text{for} \quad \begin{cases}|f| \leq B \\ 0 \ \text{if not}\end{cases} \\ \tau_h(f) = -\dfrac{1}{2\pi}\dfrac{d\phi_h(f)}{df} \end{cases} \qquad [6.1]$$

The transmission channel is assumed to be stationary here (its characteristics do not vary over the time) and linear. It only introduces linear distortions on the input signal, since $|H(f)|$ and $\tau_h(f)$ are functions of the frequency.

However, the delay $\tau_h(f)$ is, on the one hand, very negligible compared to the duration of transmission of a symbol (except for very long distance satellite transmissions) and, on the other hand, is taken into account in the choice of the sampling instants. It can therefore not be taken into account directly thereafter.

The noise $b_0(t)$ is assumed to be additive at the channel output, and, therefore, at the input of the receiver's equalization filter. It consists of the sum of the disturbing noise added to the signal during its transmission and the internal noise of the receiver reduced to its input.

The noise $b_0(t)$ is assumed to be 2nd order stationary, Gaussian, centered, of power $\sigma_{b_0}^2$, independent of the useful signal, and of bilateral power spectral density $\Gamma_{b_0}(f)$ constant on a frequency band Δf_b (equivalent energy bandwidth) given by:

$$\begin{cases} \Gamma_{b_0}(f) = \dfrac{1}{2}\Gamma_{b_0} = \Gamma_0 \quad \text{for} \quad f \in [-\Delta f_b, \Delta f_b] \\ 0 \quad \text{if not} \end{cases} \qquad [6.2]$$

The noise power $\sigma_{b_0}^2$ is given by:

$$\sigma_{b_0}^2 = \int_{-\infty}^{\infty} \Gamma_{b_0}(f)df = \Gamma_0 \int_{-\Delta f_b}^{\Delta f_b} df = 2\,\Gamma_0 \times \Delta f_b \qquad [6.3]$$

The problem dealt with in this chapter is that of optimizing all the blocks in the transmission chain in order to obtain the best estimate of the symbols $\{\hat{a}_k\}$.

The binary sequence $\{b_n\}$ is transformed by the encoder into a sequence $\{a_n\}$ where the symbols a_n are M-ary symmetric symbols:

$$a_n = \{\pm 1, \pm 3, \cdots, \pm(2m+1), \cdots, \pm(M-1)\} \qquad [6.4]$$

After filtering by the shaping filter with transfer function $X(f)$ and real impulse response $x(t)$, the digital signal transmitted $s_e(t)$ is expressed as:

$$s_e(t) = \sum_n a_n x(t - nT) \qquad [6.5]$$

The signal $x(t)$ has an amplitude V and as temporal support the interval $[0, \theta[$ in which θ denotes the period of emission of the symbols, and is not necessarily a rectangular function. In the case where $x(t)$ is a rectangle signal, we then take $\theta = T$ for the NRZ code and $\theta = T/2$ for the RZ code.

The signal emitted $s_e(t)$ is applied to the input of the physical transmission medium (cable) modeled by the low-pass filter with transfer function $H(f)$ and impulse response $h(t)$. The signal $s_r(t)$ at the output of the transmission channel (or equivalently at the input of the receiver) is of the form:

$$s_r(t) = \sum_n a_n y(t - nT) + b_0(t) \quad [6.6]$$

where $y(t)$ is the response of the physical transmission medium (cable) to the physical signal $x(t)$ (excluding noise):

$$y(t) = x(t) \otimes h(t) \quad [6.7]$$

and $\otimes$ is the convolution product.

In the frequency domain, one has:

$$Y(f) = X(f) \times H(f) \quad [6.8]$$

The input filter at the receiver, also called the "equalization filter", is intended to compensate for the harmful effects of the transmission medium. Its transfer function is $G_c(f)$ and its impulse response $g_c(t)$. It works in a frequency band totally included in the frequency band $[-\Delta f_b, \Delta f_b]$. It delivers a corrected (also called equalized) signal $s_c(t)$:

$$s_c(t) = \sum_n a_n p(t - nT) + b_1(t) \quad [6.9]$$

with:

$$p(t) = y(t) \otimes g_c(t) = x(t) \otimes h(t) \otimes g_c(t) \quad [6.10]$$

the response of the equalization filter to signal $y(t)$, that is to say, the response of the whole (cable + equalization) to the elementary pulse $x(t)$, and:

$$b_1(t) = b_0(t) \otimes g_c(t) \quad [6.11]$$

is the result of filtering the noise $b_0(t)$ by the equalization filter. It is characterized by its power spectral density $\Gamma_{b_1}(f)$ and its power $\sigma_{b_1}^2$.

In the frequency domain, one has:

$$P(f) = Y(f) \times G_c(f) = X(f) \times H(f) \times G_c(f) \qquad [6.12]$$

The signal $s_c(t)$ is sampled at regularly spaced instants t_k of the form:

$$t_k = t_0 + kT \ ; \ k = 0, 1, 2, \cdots \qquad [6.13]$$

and applied to the input of the decision system. It is then expressed by :

$$s_c(t_0 + kT) = \sum_{n=-\infty}^{\infty} a_n\, p[t_0 + (k-n)T] + b_1(t_0 + kT) \qquad [6.14]$$

Or, by isolating the term that contains the contribution of the useful symbol a_k:

$$s_c(t_0 + kT) = a_k p(t_0) + \sum_{n=-\infty, n\neq k}^{\infty} a_n\, p[t_0 + (k-n)T] + b_1(t_0 + kT) \qquad [6.15]$$

So the signal $s_c(t_0 + kT)$ is made of the sum of three terms:

– the useful signal $a_k p(t_0)$ which corresponds to the response of the system (cable + equalization) to the transmission of the elementary symbol a_k associated with this time interval;

– the sum of responses:

$$I_{m_l}(t_0 + kT) = \sum_{n=-\infty, n\neq k}^{\infty} a_n\, p[t_0 + (k-n)T] \qquad [6.16]$$

called "intersymbol interference" (ISI). It is a parasitic signal depending on all the symbols interfering with symbol a_k transmitted. The latter is the only interesting one over the time interval considered. The set of symbols interfering with the symbol a_k defines a message m_l belonging to the set $\mathcal{M}$ of possible messages of inter-symbol interference. The index $m_l \in \mathcal{M}$ in the relation [6.16] indicates this dependence. The equalization function will aim to maximize the signal $p(t_0)$ and cancel (minimize) the ISI;

– the additive noise $b_1(t_0 + kT)$, stationary, Gaussian, centered, with a power spectral density given by:

$$\Gamma_{b_1}(f) = \Gamma_{b_0}(f) \times |G_c(f)|^2 \qquad [6.17]$$

and a total power $\sigma_{b_1}^2$ given by:

$$\sigma_{b_1}^2 = \int_{-\infty}^{\infty} \Gamma_{b_1}(f) df \qquad [6.18]$$

We call Δf_c the equivalent energy bandwidth of the equalization filter $G_c(f)$. Since $\Delta f_c < \Delta f_b$, one then has:

$$\sigma_{b_1}^2 = \Gamma_0 \int_{-\Delta f_b}^{\Delta f_b} |G_c(f)|^2 df = \Gamma_0 |G_c(0)|^2 \int_{-\Delta f_c}^{\Delta f_c} df = 2\,\Gamma_0 \Delta f_c \times |G_c(0)|^2 \qquad [6.19]$$

Or again, from [6.3]:

$$\sigma_{b_1}^2 = \sigma_{b_0}^2 \times \frac{\Delta f_c}{\Delta f_b} \times |G_c(0)|^2 \qquad [6.20]$$

6.3.1. *Equivalent energy bandwidth Δf_e of a low-pass filter*

In practical applications, to facilitate the calculation, it is often interesting to replace a filter $G(f)$ of a very large frequency band (tends towards infinity) by its equivalent $G_e(f)$ (see Figure 6.2), by determining its energy bandwidth Δf_e, with the constraint of having the same gain at zero frequency:

$$|G_e(0)| = |G(0)|$$

In energy terms, one has:

$$\int_{-\Delta f_e}^{\Delta f_e} |G_e(f)|^2 df = \int_{-\infty}^{\infty} |G(f)|^2\, df = 2\Delta f_e |G(0)|^2 \qquad [6.21]$$

hence:

$$\Delta f_e = \frac{1}{2} \int_{-\infty}^{\infty} \left|\frac{G(f)}{G(0)}\right|^2 df \qquad [6.22]$$

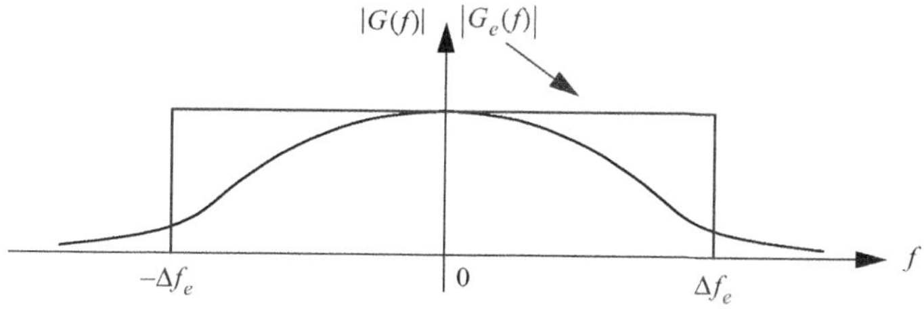

Figure 6.2. *Equivalent energy bandwidth Δf_e of a low-pass filter. For a color version of this figure, see www.iste.co.uk/assad/digital1.zip*

6.4. Characterization of the intersymbol interference: eye pattern

Let's first explain the physical phenomenon of ISI, illustrated in Figure 6.3.

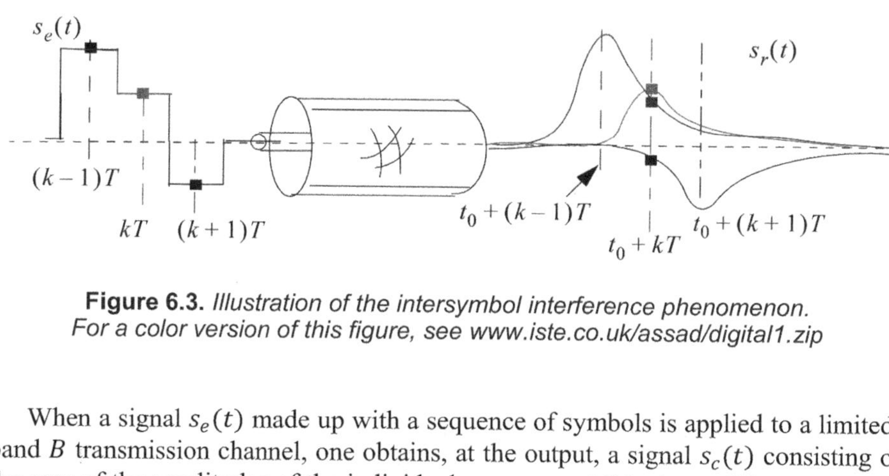

Figure 6.3. *Illustration of the intersymbol interference phenomenon. For a color version of this figure, see www.iste.co.uk/assad/digital1.zip*

When a signal $s_e(t)$ made up with a sequence of symbols is applied to a limited-band B transmission channel, one obtains, at the output, a signal $s_c(t)$ consisting of the sum of the amplitudes of the individual responses at this time t. The response to each input pulse (here, rectangular and of width T) is, as shown in Figure 6.3, of an elongated bell shape of width λT with $\lambda \gg 1$.

A significant widening of the response means that the bandwidth B of the channel is small compared to the symbol rate $D_s = 1/T$, expressed in symbols/second, of the input signal and vice versa.

The enlargement leads to the overlapping of pulse responses which, consequently, induces the phenomenon of intersymbol interference. The importance of this phenomenon depends on the number and amplitudes of the neighboring pulse responses which are added at the instant t_k to the response to the transmitted symbol

a_k. The effect of ISI can be reduced by increasing the bandwidth of the channel, or by reducing the symbol transmission rate.

6.4.1. *Eye pattern*

The eye diagram, called this because of its resemblance to the shape of a human eye, is the figure obtained by superimposing all traces or realizations of the signal $s_c(t)$ noiseless. It is periodic, of period T.

The channel is of limited bandwidth B, let us then suppose that the signal $p(t)$ extends over the interval $[-L_1T,\ L_2T]$ with L_1 and L_2 integers, and is zero or negligible outside this interval. Let us examine the set of ISI values contained in the signal $s_c(t)$ over the interval $[0,\ T[$ for all the realizations of the sequence $\{a_n\}$.

On this interval, the signal $s_c(t)$ is expressed (remember that it is assumed to be noise-free):

$$s_c(t) = \sum_{n=-L_1}^{L_2} a_n p(t - nT) \qquad 0 \leq t < T \tag{6.23}$$

The possible number of traces of the signal $s_c(t)$ is then equal to $M^{L_1+L_2+1}$, and $2^{L_1+L_2+1}$ in the case of binary symbols.

The superposition of all these traces constitutes the eye pattern associated with the signal $s_c(t)$ on the interval $[0,\ T[$. This is, of course, valid for any interval $[t_1, t_1 + T[$ where t_1 is an arbitrary instant.

The eye pattern can be displayed on an oscilloscope. The oscilloscope time base is synchronized to the sampling rate of $1/T$ frequency, so that at the instants $t_0 + kT$, the spot always occupies the same horizontal position on the screen. If the scanning velocity of the oscilloscope is sufficiently high compared to the persistence of the cathode ray tube or, even better, if the oscilloscope has a trace memory, a large number of traces of $s_c(t)$ are visible simultaneously on the screen.

The eye pattern allows direct measurement of the ISI at any time, and, therefore, an assessment of the quality of a transmission link. It shows the importance of the distortion of the digital signal, and allows us to quickly and approximately adjust the equalization function.

In Figure 6.4 we show all the characteristics that an eye pattern can have in the case of binary symbols. In the general case of M-ary symbols, we will have $(M-1)$ eyes.

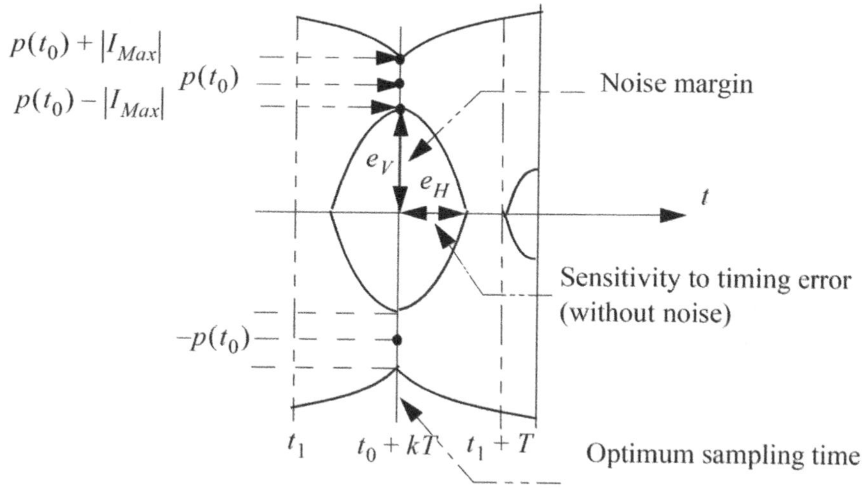

Figure 6.4. *Characteristics of the eye diagram: case of binary symbols* $a_k = \pm 1$

Consider the case where the ISI leads to decision errors in the absence of noise $b_1(t)$. As the optimal decision thresholds without ISI (see formula [6.71]) are distant from $2p(t_0)$, errors due to the ISI may occur if the absolute value of the ISI is greater than $p(t_0)$, that is to say:

$$Max_{\{a_n\}}\left|\sum_{n\neq k} a_n p[t_0 + (k-n)]T\right| > p(t_0) \qquad [6.24]$$

However, the largest absolute value of a_n is $(M-1)$, which causes the maximum distortion brought by the ISI to be:

$$I_{m_l Max}(t_0 + kT) = (M-1)\left|\sum_{n\neq k} p[t_0 + (k-n)]T\right| > p(t_0) \qquad [6.25]$$

(denoted as $|I_{Max}|$ in Figure 6.4).

When the relation [6.25] is verified, the eye pattern is completely closed at the sampling instants t_k, and decision errors are obvious, even in the absence of noise.

The distance e_V "vertical half-opening of the eye" represents the amplitude of noise corresponding to the threshold of appearance of the errors. In the absence of noise, it is sufficient that the vertical opening of the eye is greater than zero, so that it is possible to distinguish, without error, the values of two neighboring symbols.

The “horizontal half-opening of the eye” e_H represents the maximum tolerable offset of the sampling instant in the absence of noise. In the presence of noise, this tolerance is reduced.

We give two examples of eye diagrams in Figure 6.5: the first (Figure 6.5(a)) concerns the case of binary transmission $M = 2$, for a signal-to-noise ratio of 15 dB. The second (Figure 6.5(b)) concerns an M-ary transmission with $M = 4$, for a signal-to-noise ratio of 30 dB.

In both cases, the diagrams are represented for a *roll-off factor* $\alpha = 0.6$ (the *roll-off factor* coefficient will be defined precisely later in the text).

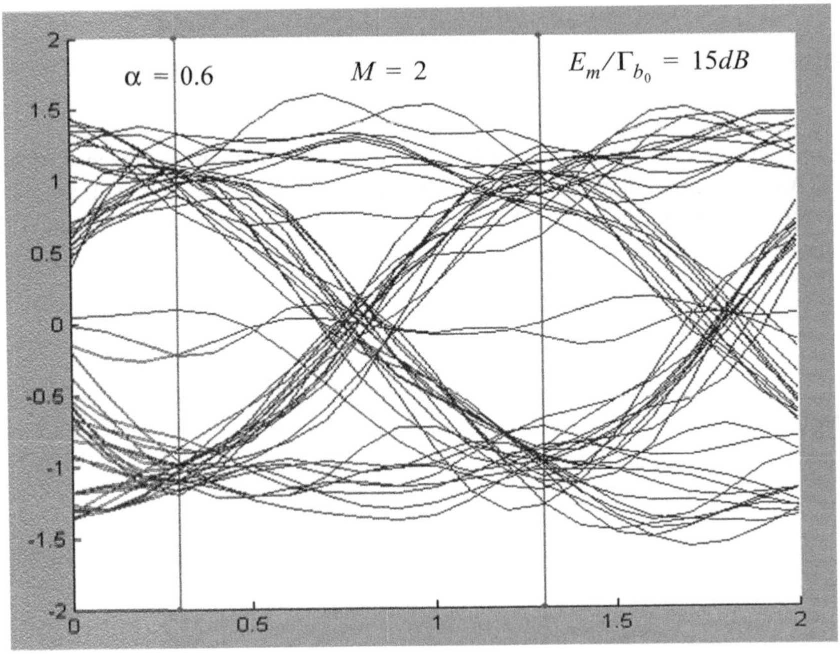

Figure 6.5(a). *Examples of an eye pattern. For a color version of this figure, see www.iste.co.uk/assad/digital1.zip*

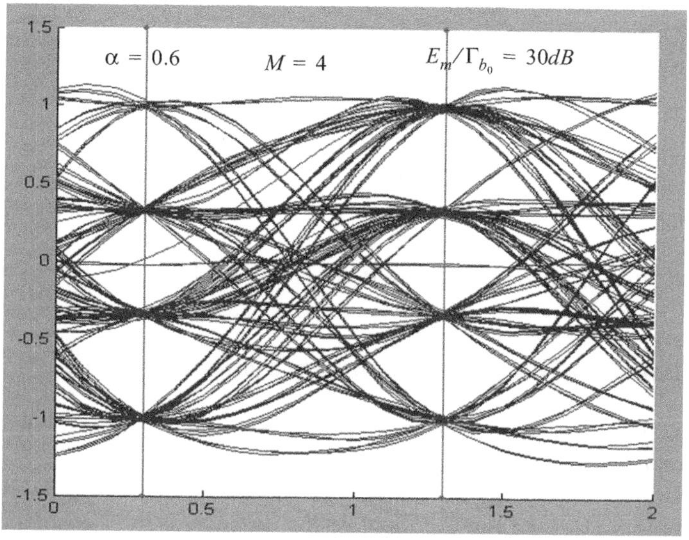

Figure 6.5(b). *Examples of an eye pattern (following). For a color version of this figure, see www.iste.co.uk/assad/digital1.zip*

6.5. Probability of error P_e

Let us first place ourselves in the practical case where the signals $p[t_0 + (k-n)T]$ are null or negligible outside the interval $[-L_1T,\ L_2T]$.

The expression of the signal obtained after equalization and sampling becomes:

$$s_c(t_0 + kT) = \sum_{n=k-L_1}^{k+L_2} a_n p[t_0 + (k-n)T] + b_1(t_0 + kT) \qquad [6.26]$$

Since $(L_1 + L_2 + 1)$ symbols a_n contribute to the response $s_c(t_0 + kT)$, the receiver is likely to receive $M^{(L_1+L_2+1)}$, different signals, each of duration $(L_1 + L_2 + 1)$. The decision could be made from $M^{(L_1+L_2+1)}$ matched filters but its complexity increases exponentially with the number $(L_1 + L_2 + 1)$.

We are interested in the decision concerning the information symbol a_k (of rank k). The signal on which the decision is performed as to the value of the information symbol a_k transmitted is:

$$s_c(t_0 + kT) = a_k p(t_0) + I_{m_l}(t_0 + kT) + b_1(t_0 + kT) \qquad [6.27]$$

with:

$$I_{m_l}(t_0 + kT) = \sum_{n=k-L_1, n\neq k}^{k+L_2} a_n p[t_0 + (k-n)T] \qquad [6.28]$$

So one has $M^L = M^{(L_1+L_2)}$ messages m_l interfering with symbol a_k, where:

$$m_l = \{a_{k-L_1}, \cdots, a_{k-1}, a_{k+1}, \cdots, a_{k+L_2}\} \text{ and } l \in [1, \cdots, M^L] \qquad [6.29]$$

6.5.1. *Probability of error: case of binary symbols* $a_k = \pm 1$

We must first note that when the symbols are binary, then we have $T = T_b$, where T_b is the duration of a transmitted binary symbol.

The sample $s_c(t_0 + kT)$ at the input of the decision system is compared with a single threshold μ_0, and a decision $\hat{a}_k = d[s_c(t_0 + kT)]$ concerning the value of the symbol a_k is taken, according to the following rule:

$$s_c(t_0 + kT) \geq \mu_0 \text{ then } \hat{a}_k = 1 \;;\quad s_c(t_0 + kT) < \mu_0 \text{ then } \hat{a}_k = -1 \qquad [6.30]$$

Two types of wrong decision are possible: take the decision $\hat{a}_k = 1$ knowing that $a_k = -1$, or vice versa.

Hence the two conditional probabilities of wrong decision:

$$P_{e_{-1}} = P_{e(1/-1, m_l)} = Pr\{\hat{a}_k = 1/a_k = -1, m_l\} \qquad [6.31]$$

$$P_{e_1} = P_{e(-1/1, m_l)} = Pr\{\hat{a}_k = -1/a_k = 1, m_l\} \qquad [6.32]$$

If $p_i = Pr\{a_k = i\}$ with $i = [-1, 1]$ is the probability to transmit the symbol $a_k = i$, and $p_{m_l} = Pr\{m_l\}$ is the probability to have the interfering message m_l, then, in the case of independent emitted symbols, the probability of error at time t_k is expressed by:

$$P_e = p_{-1}\left[\sum_{l=1}^{2^L} p_{m_l} \times P_{e_{-1}}\right] + p_1\left[\sum_{l=1}^{2^L} p_{m_l} \times P_{e_1}\right] \qquad [6.33]$$

6.5.1.1. *Calculation of conditional probabilities of erroneous decision*

The signal sample $s_c(t_0 + kT)$ in the relation [6.27] can then be considered as the realization of a random Gaussian variable of mean value:

$$a_k p(t_0) + I_{m_l}(t_0 + kT)$$

and variance:

$$\sigma_{b_1}^2 = E[b_1^2(t_0 + kT)] \quad [6.34]$$

The reasoning and the calculation which follow are based on the fact that the probability of a random variable can be calculated starting from the knowledge of its density of probability and its support, indeed:

$$Pr\{\alpha \le b < \beta\} = \int_{\alpha}^{\beta} p(b)\, db \quad [6.35]$$

The noise $b_1(t_0 + kT)$ is written:

$$b_1(t_0 + kT) = s_c(t_0 + kT) - a_k p(t_0) - I_{m_l}(t_0 + kT) \quad [6.36]$$

In general, the sample $s_c(t_0 + kT)$ belongs to an interval $[c, d[$ depending on the decision thresholds. One can then write:

$$b_1(t_0 + kT) \in \left[c - a_k p(t_0) - I_{m_l}(t_0 + kT), d - a_k p(t_0) - I_{m_l}(t_0 + kT)\right[\quad [6.37]$$

The decision rule [6.30] is set as follows:

1) One decides that:

$$\hat{a}_k = 1 \text{ if } s_c(t_0 + kT) \ge \mu_0 \;\rightarrow s_c(t_0 + kT) \in [\mu_0, \infty[\quad [6.38]$$

hence:

$$b_1(t_0 + kT) \in \left[\mu_0 - a_k p(t_0) - I_{m_l}(t_0 + kT), \infty\right[\quad [6.39]$$

and:

$$P_{e_{-1}} = Pr\{\hat{a}_k = 1/a_k = -1, m_l\} = Pr\{\mu_0 \le s_c(t_0 + kT) < \infty/a_k = -1, m_l\}$$

$$P_{e_{-1}} = Pr\left\{\mu_0 + p(t_0) - I_{m_l}(t_0 + kT) \le b_1(t_0 + kT) < \infty\right\} \quad [6.40]$$

so:

$$P_{e_{-1}} = \frac{1}{\sigma_{b_1}\sqrt{2\pi}} \int_{\mu_0+p(t_0)-I_{m_l}(t_0+kT)}^{\infty} \exp\left[\frac{-1}{2}\frac{b_1^2}{\sigma_{b_1}^2}\right] db_1 \quad [6.41]$$

or:

$$P_{e_{-1}} = \frac{1}{2}\operatorname{erfc}\left[\frac{\mu_0 + p(t_0) - I_{m_l}(t_0 + kT)}{\sigma_{b_1}\sqrt{2}}\right] \quad [6.42]$$

2) In the same way, one decides that:

$$\hat{a}_k = -1 \text{ if } s_c(t_0 + kT) < \mu_0 \ \rightarrow s_c(t_0 + kT) \in [-\infty, \mu_0[\quad [6.43]$$

hence:

$$b_1(t_0 + kT) \in \left[-\infty, \mu_0 - a_k p(t_0) - I_{m_l}(t_0 + kT)\right[\quad [6.44]$$

and:

$$P_{e_1} = Pr\{\hat{a}_k = -1/a_k = 1, m_l\}$$

$$= Pr\{-\infty \leq s_c(t_0 + kT) < \mu_0/a_k = 1, m_l\}$$

$$P_{e_1} = Pr\left\{-\infty \leq b_1(t_0 + kT) < \mu_0 - p(t_0) - I_{m_l}(t_0 + kT)\right\} \quad [6.45]$$

so:

$$P_{e_1} = \frac{1}{\sigma_{b_1}\sqrt{2\pi}} \int_{-\infty}^{\mu_0-\left[p(t_0)+I_{m_l}(t_0+kT)\right]} \exp\left[\frac{-1}{2}\frac{b_1^2}{\sigma_{b_1}^2}\right] db_1 \quad [6.46]$$

or:

$$P_{e_1} = \frac{1}{2}\operatorname{erfc}\left[\frac{-\mu_0 + p(t_0) + I_{m_l}(t_0 + kT)}{\sigma_{b_1}\sqrt{2}}\right] \quad [6.47]$$

Recall that the probability density of the noise b_1 is given by:

$$p(b_1) = \frac{1}{\sigma_{b_1}\sqrt{2\pi}} \exp\left[\frac{-1}{2}\frac{b_1^2}{\sigma_{b_1}^2}\right] \quad [6.48]$$

and:

$$\begin{cases} \operatorname{erf}(x) = \dfrac{2}{\sqrt{\pi}} \displaystyle\int_0^x \exp(-u^2)du \ ; \ \operatorname{erf}(-x) = -\operatorname{erf}(x) \\ \operatorname{erfc}(x) = 1 - \operatorname{erf}(x) = \dfrac{2}{\sqrt{\pi}} \displaystyle\int_x^\infty \exp(-u^2)du \ ; \ \operatorname{erfc}(-x) = 2 - \operatorname{erfc}(x) \end{cases} \qquad [6.49]$$

Functions $\operatorname{erf}(x)$ and $\operatorname{erfc}(x)$ are called the “error function” and the “complementary error function”.

Some authors prefer to use the function $Q(x)$ defined by:

$$Q(x) = \frac{1}{\sqrt{2\pi}} \int_x^\infty \exp\left(\frac{-u^2}{2}\right) du \qquad [6.50]$$

and, in relation with function $\operatorname{erfc}(x)$, by:

$$\begin{cases} \operatorname{erfc}(x) = 2Q(x\sqrt{2}) \\ Q(x) = \dfrac{1}{2}\operatorname{erfc}\left(\dfrac{x}{\sqrt{2}}\right) \end{cases} \qquad [6.51]$$

6.5.1.2. *Connection with the detection theory*

There is a direct connection with the statistical detection theory. Because the false alarm probability P_{FA} (false positive, FP) can be directly deduced with what was previously described:

$$\begin{aligned} &Pr\{\hat{a}_k = 1/a_k = -1\} = Pr\{\mu_0 \leq s_c < \infty/a_k = -1\} \\ &= \int_{\mu_0}^\infty p(s_c/a_k = -1)ds_c = P_{FA} \end{aligned} \qquad [6.52]$$

It is the same for the probability of false non detection P_{ND} (false negative, FN):

$$\begin{aligned} &Pr\{\hat{a}_k = -1/a_k = 1\} = Pr\{-\infty \leq s_c < \mu_0/a_k = 1\} \\ &= \int_{-\infty}^{\mu_0} p(s_c/a_k = 1)ds_c = P_{ND} \end{aligned} \qquad [6.53]$$

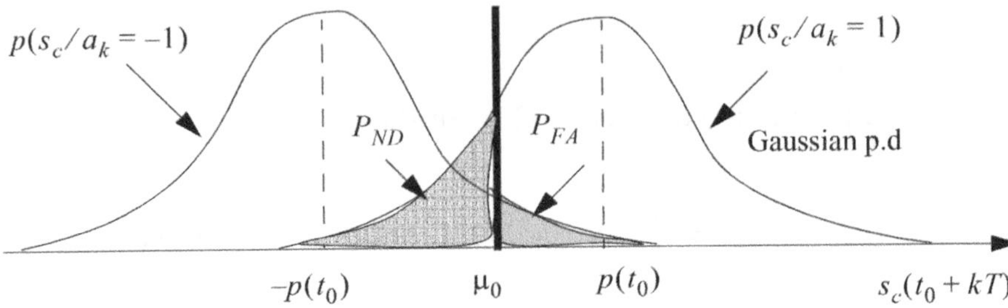

Figure 6.6. *Link with the detection theory*

We can then define the two following probabilities:

– the probability of decision P_D:

$$P_D = 1 - P_{ND} \quad [6.54]$$

– the probability of correct decision P_C:

$$P_C = 1 - P_e \quad [6.55]$$

with:

$$P_e = p_{-1}P_{FA} + p_1 P_{ND} \quad [6.56]$$

6.5.1.3. *Optimal decision threshold*

The optimal decision threshold that minimizes the probability of error P_e is the one that makes null the derivative of P_e with μ_0 ($I_{m_l}(t_0 + kT) = 0$).

Knowing that:

$$\frac{d}{dx}\left[\int_{\alpha}^{x} f(t)dt\right] = f(x) \quad [6.57]$$

then:

$$\frac{\partial P_e}{\partial \mu_0} = \left[p_1 \exp\left\{-\frac{[\mu_0 - p(t_0)]^2}{2\sigma_{b_1}^2}\right\} - p_{-1}\exp\left\{-\frac{[\mu_0 + p(t_0)]^2}{2\sigma_{b_1}^2}\right\}\right] = 0 \quad [6.58]$$

This leads to the expression of the optimal threshold, denoted μ_{0opt}, given by:

$$\mu_{0opt} = \frac{\sigma_{b_1}^2}{2p(t_0)} \log_e \left[\frac{p_{-1}}{p_1}\right] \qquad [6.59]$$

The optimal threshold only depends on the probability distribution $\{p_1, p_{-1}\}$. For example, if $p_1 > p_{-1}$, the threshold moves to negative values, so as to favor the decision $\hat{a}_k = 1$, that is to say, the most likely *a priori* binary symbol.

In the case where the binary symbols transmitted are equiprobable, $p_1 = p_{-1} = 1/2$, then the optimal threshold μ_{0opt} is zero, and the decision system tests only the sign of the sample $s_c(t_0 + kT)$.

6.5.1.4. *Probability of error in the simple case of equiprobability of symbols a_k*

In this case, one has:

$$p_1 = p_{-1} = 1/2 \;\; \rightarrow \mu_{0opt} = 0 \;\; ; \;\; \text{and} \;\; p_{m_l} = 1/2^L \qquad [6.60]$$

Hence the probability of error P_e:

$$P_e = \frac{1}{2^{L+1}} \sum_{l=1}^{2^L} \left[P_{e_{-1}} + P_{e_1}\right] \qquad [6.61]$$

6.5.2. *Probability of error: case of binary RZ code*

By adopting the same approach as above, the probability of error for a binary RZ code, according to the relation [6.33], is expressed by:

$$P_e = p_0 \left[\sum_{l=1}^{2^L} p_{m_l} \times P_{e_0}\right] + p_1 \left[\sum_{l=1}^{2^L} p_{m_l} \times P_{e_1}\right] \qquad [6.62]$$

with:

$$P_{e_0} = P_{e(1/0,m_l)} = \frac{1}{\sigma_{b_1}\sqrt{2\pi}} \int_{\mu_0 - I_{m_l}(t_0+kT)}^{\infty} \exp\left[-\frac{1}{2}\frac{b_1^2}{\sigma_{b_1}^2}\right] db_1 \qquad [6.63]$$

$$P_{e_1} = P_{e(0/1,m_l)} = \frac{1}{\sigma_{b_1}\sqrt{2\pi}} \int_{-\infty}^{\mu_0 - p(t_0) - I_{m_l}(t_0+kT)} \exp\left[-\frac{1}{2}\frac{b_1^2}{\sigma_{b_1}^2}\right] db_1$$

and:

$$p_0 = Pr\{a_k = 0\} \qquad [6.64]$$

The optimal threshold is obtained by calculating the threshold value which cancels the derivative of the probability of error in optimal reception:

$$\frac{\partial P_e}{\partial \mu_0} = \left[p_1 \exp\left\{-\frac{[\mu_0 - p(t_0)]^2}{2\sigma_{b_1}^2}\right\} - p_0 \exp\left\{-\frac{[\mu_0]^2}{2\sigma_{b_1}^2}\right\}\right] = 0 \qquad [6.65]$$

hence:

$$\mu_{0opt} = \frac{p(t_0)}{2} + \frac{\sigma_{b_1}^2}{p(t_0)} \log_e \left[\frac{p_0}{p_1}\right] \qquad [6.66]$$

In the case of equiprobable symbols, the optimal threshold becomes:

$$\mu_{0opt} = \frac{p(t_0)}{2} \qquad [6.67]$$

6.5.2.1. *Probability of error in the case of equiprobable symbols*

In this case:

$$p_1 = p_0 = 1/2 \ \rightarrow \mu_0 = \mu_{0opt} = \frac{p(t_0)}{2} \ ; \ \text{and} \ p_{m_l} = 1/2^L \qquad [6.68]$$

and then:

$$P_e = \frac{1}{2^{L+1}} \sum_{l=1}^{2^L} [P_{e_0} + P_{e_1}] \qquad [6.69]$$

6.5.3. *Probability of error: general case of M-ary symbols*

Let us now consider the general case where the symbols are M-ary and take the following values:

$$a_n = \{\pm 1, \pm 3, \cdots, \pm(2m+1), \cdots, \pm(M-1)\} \text{ and } M = 2^k \qquad [6.70]$$

The waveform $x(t)$ is of duration $T = kT_b$. The structure of the receiver remains unchanged. It is composed of a reception filter (impulse response $g_c(t)$) followed by a sampler and a threshold comparator system with $(M-1)$ thresholds μ_m. The thresholds μ_m, in the case of optimal reception and equiprobable symbols, are optimum and located in the center of each interval, delimited by two consecutive values of the sample $a_k p(t_0)$:

$$\mu_m = \mu_{mopt} = 2m \times p(t_0) \qquad [6.71]$$

with:

$$m \in \left\{\left(\frac{-M}{2}+1\right), \cdots, 0, \cdots, \left(\frac{M}{2}-1\right)\right\} \qquad [6.72]$$

where m is a relative integer.

In Figure 6.7 we show the different possible values of the sample $a_k p(t_0)$ and the position of the optimal thresholds.

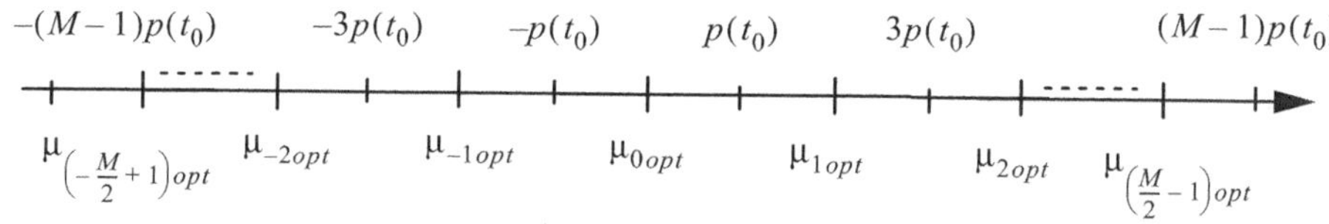

Figure 6.7. *Sample values $a_k p(t_0)$ and optimal threshold values (M-ary symbols)*

6.5.3.1. *Decision rules (symbols $a_k : \hat{a}_k$)*

6.5.3.1.1. Case of intermediate values $2m+1$, where $m \neq -\frac{M}{2}$ and $m \neq \left(\frac{M}{2}-1\right)$

The decision on symbol a_k is carried out by comparing the sample $s_c(t_0 + kT)$ to the different thresholds μ_m, and the decision $\hat{a}_k = (2m+1)$ is taken as soon as $s_c(t_0 + kT)$ lies between the two consecutive thresholds μ_m and μ_{m+1}:

$$\hat{a}_k = (2m+1) \text{ if } \mu_m \leq s_c(t_0 + kT) < \mu_{m+1}$$
$$\text{for } m \neq -\frac{M}{2} \text{ and } m \neq \left(\frac{M}{2} - 1\right) \qquad [6.73]$$

6.5.3.1.2. Case of extreme values $\pm(M-1)$

To decide on extreme values $\pm(M-1)$, the sample $s_c(t_0 + kT)$ is compared to a single threshold in each case:

$$\hat{a}_k = (M-1) \Leftrightarrow s_c(t_0 + kT) \geq \mu_{\frac{M}{2}-1} \Rightarrow \mu_{\frac{M}{2}-1} \leq s_c(t_0 + kT) < \infty$$
$$\hat{a}_k = -(M-1) \Leftrightarrow s_c(t_0 + kT) < \mu_{-\frac{M}{2}+1} \Rightarrow -\infty \leq s_c(t_0 + kT) < \mu_{-\frac{M}{2}+1} \qquad [6.74]$$

Denoting by p_{2i+1} the probability of emission of the symbols a_k:

$$p_{2i+1} = Pr\{a_k = 2i+1\} \qquad [6.75]$$

and by $P_{e_{2i+1}}$ the conditional probability of error which, conditionally to each value of the symbol a_k transmitted and of all the messages m_l interfering with it, is written:

$$P_{e_{2i+1}} = P_{e(2m+1/2i+1, m_l)_{m \neq i}} = Pr\{\hat{a}_k = 2m+1/a_k = 2i+1, m_l\}_{m \neq i} \qquad [6.76]$$

with:

$$i, m \in \left\{\frac{-M}{2}, \cdots, 0, \cdots, \left(\frac{M}{2} - 2\right)\right\} \qquad [6.77]$$

The total average probability of error (wrong decision) is then expressed by:

$$P_e = \sum_{i=-\frac{M}{2}}^{\frac{M}{2}-1} p_{2i+1} \left[\sum_{l=1}^{M^l} p_{m_l} \sum_{m=-\frac{M}{2}}^{\frac{M}{2}-1} Pr\{\hat{a}_k = 2m+1/a_k = 2i+1, m_l\}_{m \neq i}\right] \qquad [6.78]$$

with:

$$p_{m_l} = Pr\{m_l\} \qquad [6.79]$$

the probability of having the interfering message m_l.

The expression [6.78] can be also written as:

$$P_e = \sum_{i=1}^{M} p_i \left[\sum_{l=1}^{M^l} p_{m_l} \sum_{j=1}^{M} Pr\{\hat{a}_k = a_j / a_k = a_i, m_l\}_{j \neq i} \right] \quad [6.80]$$

In the case where $M > 4$ and non independent and non equiprobable symbols a_k are transmitted, the expression [6.78] of the probability of error is, in general, difficult to calculate analytically. In such cases, the computation is carried out by computer.

If the symbols a_k are independent and equiprobable, then one has:

$$p_{2i+1} = \frac{1}{M} \; ; \; p_{m_l} = \frac{1}{M^L} \quad [6.81]$$

The expression [6.78] of the error probability becomes:

$$P_e = \frac{1}{M^{L+1}} = \sum_{i=-\frac{M}{2}}^{\frac{M}{2}-1} \left[\sum_{l=1}^{M^l} \sum_{m=-\frac{M}{2}}^{\frac{M}{2}-1} Pr\{\hat{a}_k = 2m+1 / a_k = 2i+1, m_l\}_{m \neq i} \right] \quad [6.82]$$

6.5.3.2. *Calculation of the conditional probabilities of error in the case of M-ary symbols (M is even)*

Using the decision rule given by the relations [6.73] and [6.74], we can calculate the conditional probabilities of errors $P_{e_{2i+1}}$ as follows (we develop the same approach as in the binary case).

In the receiver, the noise at the decision instant is written:

$$b_1(t_0 + kT) = s_c(t_0 + kT) - a_k p(t_0) - I_{m_l}(t_0 + kT) \quad [6.83]$$

If the signal $s_c(t_0 + kT)$ at the input of the decision system belongs to the interval $[c, d[$, consequently the amplitude of the noise $b_1(t_0 + kT)$ belongs to the following interval:

$$b_1(t_0 + kT) \in \left[c - a_k p(t_0) - I_{m_l}(t_0 + kT), d - a_k p(t_0) - I_{m_l}(t_0 + kT)\right[\quad [6.84]$$

The decision system operates according to the rules [6.73] and [6.74], so as follows.

6.5.3.2.1. Case of intermediate values of decision

It is decided that:

$$\hat{a}_k = (2m+1) \text{ if } \mu_m \leq s_c(t_0 + kT) < \mu_{m+1} \text{ with } m \neq \frac{-M}{2} \text{ and } m \neq \frac{M}{2} - 1 \quad [6.85]$$

hence:

$$[c, d[= [\mu_m, \mu_{m+1}[\text{ and } \mu_m = \mu_{mopt} = 2m \times p(t_0) \quad [6.86]$$

then:

$$b_1(t_0 + kT) \in [\mu_m - a_k p(t_0) - I_{m_l}(t_0 + kT), \mu_{m+1} - a_k p(t_0) - I_{m_l}(t_0 + kT)[\quad [6.87]$$

Given $a_k = (2i+1)$, the conditional probability of error $P_{e_{2i+1}}$ is written:

$$P_{e_{2i+1}} = Pr\{\hat{a}_k = 2m + 1/a_k = 2i + 1, m_l\}$$

$$P_{e_{2i+1}} = Pr\{\mu_m \leq s_c(t_0 + kT) < \mu_{m+1}/a_k = 2i + 1, m_l\}$$

$$= Pr\{\mu_m - a_k p(t_0) - I_{m_l}(t_0 + kT) \leq b_1(t_0 + kT) \quad [6.88]$$

$$< \mu_{m+1} - a_k p(t_0) - I_{m_l}(t_0 + kT)\}$$

where:

$$m \neq i \; ; \; m \neq \frac{-M}{2} \text{ and } m \neq \frac{M}{2} - 1 \quad [6.89]$$

so:

$$P_{e_{2i+1}} = \frac{1}{\sigma_{b_1}\sqrt{2\pi}} \left[\int_{2mp(t_0)-(2i+1)p(t_0)-I_{m_l}(t_0+kT)}^{2(m+1)p(t_0)-(2i+1)p(t_0)-I_{m_l}(t_0+kT)} \exp\left[\frac{-1}{2}\frac{b_1^2}{\sigma_{b_1}^2}\right] db_1 \right] \quad [6.90]$$

or:

$$P_{e_{2i+1}} = \frac{1}{2}\{\mathrm{erfc}\left[\frac{2mp(t_0) - (2i+1)p(t_0) - I_{m_l}(t_0+kT)}{\sigma_{b_1}\sqrt{2}}\right]$$
$$-\mathrm{erfc}\left[\frac{2(m+1)p(t_0) - (2i+1)p(t_0) - I_{m_l}(t_0+kT)}{\sigma_{b_1}\sqrt{2}}\right]\} \quad [6.91]$$

6.5.3.2.2. Case of extreme value of decision $\hat{a}_k = (M-1)$

It is decided that:

$$\hat{a}_k = (M-1) \quad \text{if} \quad s_c(t_0+kT) \geq \mu_{\frac{M}{2}-1} \quad [6.92]$$

hence:

$$[c, d[= \left[\mu_{\frac{M}{2}-1}, \infty\right[\quad [6.93]$$

then:

$$b_1(t_0+kT) \in \left[\mu_{\frac{M}{2}-1} - a_k p(t_0) - I_{m_l}(t_0+kT), \infty\right[\quad [6.94]$$

with:

$$\mu_{\frac{M}{2}-1} = 2\left[\frac{M}{2} - 1\right]p(t_0) \quad [6.95]$$

Given $a_k = (2i+1) \neq (M-1)$, the conditional probability of error $P_{e_{2i+1}}$ is written:

$$P_{e_{2i+1}} = Pr\{\hat{a}_k = (M-1)/a_k = 2i+1 \neq (M-1), m_l\}$$
$$= Pr\left\{\mu_{\frac{M}{2}-1} \leq s_c(t_0+kT) < \infty/a_k = 2i+1 \neq (M-1), m_l\right\}$$
$$= Pr\left\{\mu_{\frac{M}{2}-1} - a_k p(t_0) - I_{m_l}(t_0+kT) \leq b_1(t_0+kT) < \infty\right\}$$

$$P_{e_{2i+1}} = \frac{1}{\sigma_{b_1}\sqrt{2\pi}}\left[\int_{(M-2)p(t_0)-(2i+1)p(t_0)-I_{m_l}(t_0+kT)}^{\infty} exp\left[\frac{-1}{2}\frac{b_1^2}{\sigma_{b_1}^2}\right]db_1\right] \quad [6.96]$$

with:

$$(2i+1) \neq M-1 \qquad [6.97]$$

or:

$$P_{e_{2i+1}} = \frac{1}{2}\operatorname{erfc}\left[\frac{(M-2)p(t_0) - (2i+1)p(t_0) - I_{m_l}(t_0+kT)}{\sigma_{b_1}\sqrt{2}}\right] \qquad [6.98]$$

6.5.3.2.3. Case of the extreme value of decision $\hat{a}_k = -(M-1)$

It is decided that:

$$\hat{a}_k = -(M-1) \ \text{ if } \ s_c(t_0+kT) < \mu_{-\frac{M}{2}+1} \,] \qquad [6.99]$$

hence:

$$[c,d[= \left[-\infty, \mu_{-\frac{M}{2}+1}\right[\qquad [6.100]$$

then:

$$b_1(t_0+kT) \in \left[-\infty, \mu_{-\frac{M}{2}+1} - a_k p(t_0) - I_{m_l}(t_0+kT)\right[\qquad [6.101]$$

with:

$$\mu_{-\frac{M}{2}+1} = 2\left[-\frac{M}{2}+1\right]p(t_0) \qquad [6.102]$$

Given $a_k = (2i+1) \neq -(M-1)$, the conditional probability of error $P_{e_{2i+1}}$ is written:

$$\begin{aligned} P_{e_{2i+1}} &= Pr\{\hat{a}_k = -(M-1)/a_k = 2i+1 \neq -(M-1), m_l\} \\ &= Pr\left\{-\infty \leq s_c(t_0+kT) < \mu_{-\frac{M}{2}+1}/a_k = 2i+1 \neq -(M-1), m_l\right\} \\ &= Pr\left\{-\infty \leq b_1(t_0+kT) < \mu_{-\frac{M}{2}+1} - a_k p(t_0) - I_{m_l}(t_0+kT)\right\} \end{aligned}$$

$$P_{e_{2i+1}} = \frac{1}{\sigma_{b_1}\sqrt{2\pi}} \left[\int_{-\infty}^{-(M-2)p(t_0)-(2i+1)p(t_0)-I_{m_l}(t_0+kT)} \exp\left[\frac{-1}{2}\frac{b_1^2}{\sigma_{b_1}^2}\right] db_1 \right] \quad [6.103]$$

with:

$$2i + 1 \neq -(M-1) \quad [6.104]$$

or:

$$P_{e_{2i+1}} = \frac{1}{2}\operatorname{erfc}\left[\frac{(M-2)p(t_0) + (2i+1)p(t_0) + I_{m_l}(t_0 + kT)}{\sigma_{b_1}\sqrt{2}}\right] \quad [6.105]$$

This makes it possible to then use the relation [6.78] to calculate the total probability of error.

6.5.4. *Probability of error: case of bipolar code*

The bipolar code is a code with $M = 3$ levels such as:

$$\begin{cases} b_k = 0 \to a_k = 0 \\ b_k = 1 \to a_n = \pm 1 \text{ alternately} \end{cases} \quad [6.106]$$

with the two threshold values:

$$\mu^+ = -\mu^- = p(t_0)/2 \quad [6.107]$$

As an example, to calculate the probability of error, we will assume that $\{b_k\}$ are independent but not equiprobable, with:

$$\begin{cases} Pr\{b_k = 0\} = 2/5 \to p_0 = Pr\{a_k = 0\} = 2/5 \\ Pr\{b_k = 1\} = 3/5 \to p_{-1} = Pr\{a_k = -1\} = p_1 = Pr\{a_k = 1\} = 3/10 \end{cases} \quad [6.108]$$

Given the symbol a_k transmitted, the calculation of the conditional probabilities of the different messages of the form $m_l = (a_{k-1}, a_{k+1})$ interfering with symbol a_k are given in Tables 6.2, 6.3 and 6.4.

$\{m_l/a_k = 0\}$ $m_l = [a_{k-1}, a_{k+1}]$		$p_{(m_l/0)} = Pr\{m_l/a_k = 0\}$
0	0	$Pr\{b_{k-1} = 0, b_{k+1} = 0\} = 4/25$
0	1	$Pr\{b_{k-1} = 0, b_{k+1} = 1 \text{ and } a_{k+1} > 0\} = 6/50$
0	-1	$Pr\{b_{k-1} = 0, b_{k+1} = 1 \text{ and } a_{k+1} < 0\} = 6/50$
1	0	$Pr\{b_{k-1} = 1, b_{k+1} = 0 \text{ and } a_{k-1} > 0\} = 6/50$
1	-1	$Pr\{b_{k-1} = 1, b_{k+1} = 1 \text{ and } a_{k-1} > 0\} = 9/50$
-1	0	$Pr\{b_{k-1} = 1, b_{k+1} = 0 \text{ and } a_{k-1} < 0\} = 6/50$
-1	1	$Pr\{b_{k-1} = 1, b_{k+1} = 1 \text{ and } a_{k-1} < 0\} = 9/50$

Table 6.2. *Calculation of the conditional probabilities* $p_{(m_l/0)} = Pr\{m_l/a_k = 0\}$

$\{m_l/a_k = -1\}$ $m_l = [a_{k-1}, a_{k+1}]$		$p_{(m_l/-1)} = Pr\{m_l/a_k = -1\}$
0	0	$Pr\{b_{k-1} = 0, b_{k+1} = 0\} = 4/25$
0	1	$Pr\{b_{k-1} = 0, b_{k+1} = 1\} = 6/25$
1	0	$Pr\{b_{k-1} = 1, b_{k+1} = 0\} = 6/25$
1	1	$Pr\{b_{k-1} = 1, b_{k+1} = 1\} = 9/25$

Table 6.3. *Calculation of the conditional probabilities* $p_{(m_l/-1)} = Pr\{m_l/a_k = -1\}$

$\{m_l/a_k = 1\}$ $m_l = [a_{k-1}, a_{k+1}]$	$p_{(m_l/1)} = Pr\{m_l/a_k = 1\}$
0 0	$Pr\{b_{k-1} = 0, b_{k+1} = 0\} = 4/25$
0 -1	$Pr\{b_{k-1} = 0, b_{k+1} = 1\} = 6/25$
-1 0	$Pr\{b_{k-1} = 1, b_{k+1} = 0\} = 6/25$
-1 -1	$Pr\{b_{k-1} = 1, b_{k+1} = 1\} = 9/25$

Table 6.4. *Calculation of the conditional probabilities* $p_{(m_l/1)} = Pr\{m_l/a_k = 1\}$

In total, there are nine different messages, of probabilities:

$$p_{m_l} = \sum_{i=-1}^{1} p(i, m_l) = \sum_{i=-1}^{1} p_i \times p_{(m_l/i)} \qquad [6.109]$$

with:

$$\sum_{l=1}^{9} p_{m_l} = 1 \qquad [6.110]$$

EXAMPLE.–

$m_1 = (0,0) = \rightarrow p_{m_1} = 3/10 \times 4/25 + 2/5 \times 4/25 + 3/10 \times 4/25 = 4/25$

$m_2 = (0,1) \rightarrow p_{m_2} = 3/10 \times 6/25 + 2/5 \times 6/50 = 3/25$

etc.

The probability of error (made of 30 terms) is then:

$$P_e = \sum_{i=-1}^{1} p_i \left[\sum_{l=1}^{9} p_{(m_l/i)} \sum_{m=-1}^{1} P_{e(m/i,m_l)_{m \neq i}} \right] \qquad [6.111]$$

If only adjacent decision errors are practically possible, i.e.:

$$a_k = \pm 1 \;\rightarrow\; \hat{a}_k = 0 \;;\; a_k = 0 \;\rightarrow\; \hat{a}_k = \pm 1 \qquad [6.112]$$

then P_e will have only 22 terms.

6.6. Conditions of absence of intersymbol interference: Nyquist criteria

6.6.1. *Nyquist temporal criterion*

The principle of cancellation of the intersymbol interference implies that, after equalization, at the sampling instants $t_k = t_0 + kT$ and according to the expression [6.16], the signal $p(t)$ fulfills the following temporal condition:

$$p[t_0 + (k-n)T] = \begin{cases} p(t_0) & \text{for } n = k \\ 0 & \text{for } n \neq k \end{cases} \qquad [6.113]$$

Taking t_0 as time origin, the previous temporal condition becomes (see also Figure 6.8):

$$p[(k-n)T] = \begin{cases} p(0) & \text{for } n = k \\ 0 & \text{for } n \neq k \end{cases} \qquad [6.114]$$

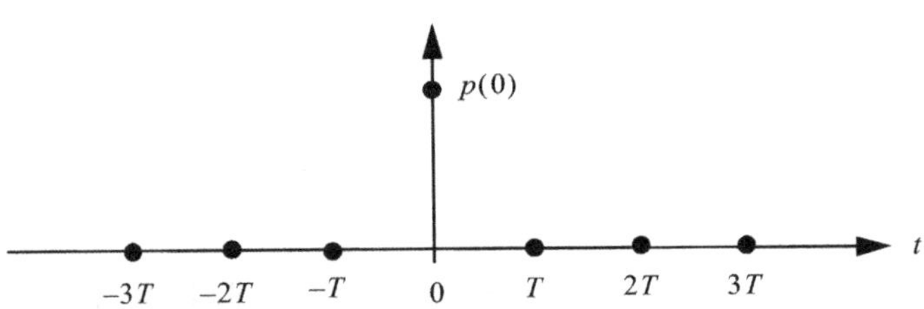

Figure 6.8. *Illustration of Nyquist temporal criterion for a null ISI*

6.6.2. *Nyquist frequency criterion*

The pulse $p(t)$ can be of any length and shape, but all of its samples $p[(k-n)T]$ with $n \neq k$ must be zero. To find the Nyquist frequency criterion, let us look at the transformation of the time condition [6.114] in the frequency domain.

For this, consider the sampled signal $p_\Delta(t)$ defined by:

$$p_\Delta(t) = p(t) \times \sum_{i=-\infty}^{\infty} \delta(t - iT) = \sum_{i=-\infty}^{\infty} p(iT)\delta(t - iT) \quad [6.115]$$

Its Fourier transform $P_\Delta(f)$ is written:

$$P_\Delta(f) = P(f) \otimes \frac{1}{T} \sum_{k=-\infty}^{\infty} \delta\left(f - \frac{k}{T}\right) = \frac{1}{T} \sum_{k=-\infty}^{\infty} P\left(f - \frac{k}{T}\right) \quad [6.116]$$

Besides, by the definition of the Fourier transform, we have:

$$P_\Delta(f) = \int_{-\infty}^{\infty} \left[p(t) \times \sum_{i=-\infty}^{\infty} \delta(t - iT)\right] \exp(-j2\pi f t) dt \quad [6.117]$$

However, the relation [6.115] is also written by using the relation [6.114]:

$$p_\Delta(t) = p(t) \times \sum_{i=-\infty}^{\infty} \delta(t - iT) \equiv p(0)\,\delta(t) \quad [6.118]$$

thus:

$$P_\Delta(f) = p(0) \int_{-\infty}^{\infty} \delta(t) \exp(-j2\pi f t) dt = p(0) \quad [6.119]$$

Finally, from [6.116] and [6.119], we deduce the Nyquist frequency criterion:

$$\sum_{k=-\infty}^{\infty} P\left(f - \frac{k}{T}\right) = T \times p(0) = \text{Constant} \quad [6.120]$$

Any function solution of [6.120] fulfills the Nyquist frequency criterion.

6.6.3. *Interpretation of the Nyquist frequency criterion*

The transmission channel (physical medium) has a bandwidth B. Its transfer function $H(f)$ satisfies:

$$H(f) = 0 \quad \text{for} \quad |f| > B \quad [6.121]$$

and, consequently, according to [6.12]:

$$P(f) = 0 \quad \text{for} \quad |f| > B \qquad [6.122]$$

We can distinguish three cases.

First case

$$\frac{1}{T} = D_s > 2B$$

The folded spectrum consists of replicas of $P(f)$ separated by $1/T$ each other, which do not overlap, and their addition cannot give a constant function (see Figure 6.9). The Nyquist frequency criterion is not verified. This means that we cannot transmit a digital signal with symbol rate D_s (symbol/s) without an ISI in a channel with a bandwidth $B < D_s/2$.

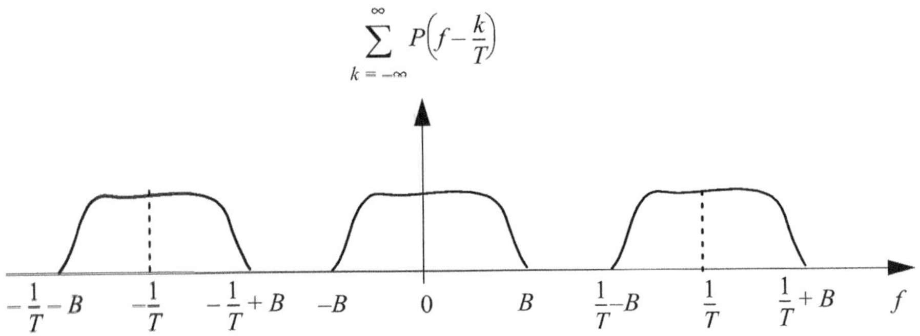

Figure 6.9. *Spectrum of* $\sum_{k=-\infty}^{\infty} P\left(f-\frac{k}{T}\right)$ *in the case where* $D_s > 2B$

Second case

$$\frac{1}{T} = D_s = 2B$$

In this case, the Nyquist criterion is only fulfilled for a given $P(f)$, denoted $N_m(f)$, (see Figures 6.10 and 6.11):

$$P(f) = N_m(f) = \begin{cases} T \times p(0) & \text{for } |f| \leq B = 1/2T \\ 0 & \text{elsewhere} \end{cases} \qquad [6.123]$$

This means that the minimum bandwidth B of the channel required to transmit a digital signal with symbol rate $D_s = 1/T$ and without ISI, is $B = 1/2T$, called the Nyquist band (by reasoning on the positive frequencies).

Otherwise, in a channel of width B, one can transmit without ISI a maximum symbol rate of:

$$D_{sMax} = 1/T = 2B \quad [6.124]$$

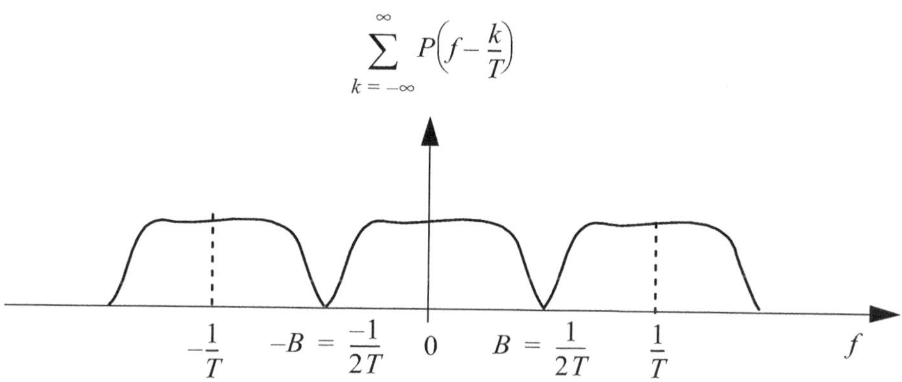

Figure 6.10. *Spectrum of $\sum_{k=-\infty}^{\infty} P\left(f - \frac{k}{T}\right)$ in the case where $D_s = 2B$*

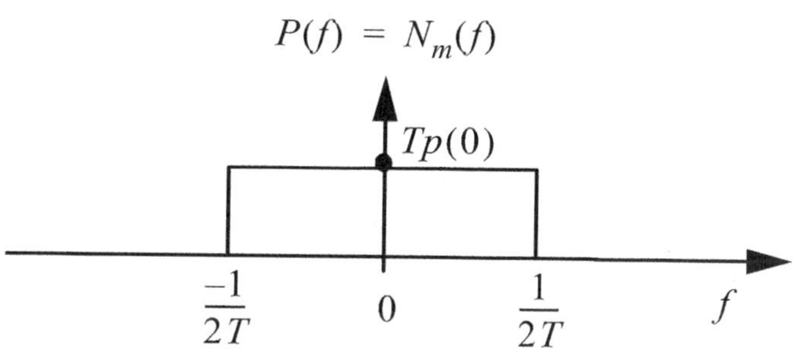

Figure 6.11. *First Nyquist frequency criterion*

This corresponds to the *first Nyquist criterion*. It should be noted that the first Nyquist criterion sets a maximum limit at the symbol rate D_s, in symbols per second, and not at the bitrate D_b, in bits per second.

Indeed, according to the following relation:

$$D_{bMax} = D_{sMax} \times \log_2 M \quad [6.125]$$

At a given D_{sMax} , the bitrate D_{bMax} can be increased by increasing the number M of levels of the signal. However, the latter is limited by the signal-to-noise ratio.

In the temporal domain, the first Nyquist criterion is translated into the following relation:

$$p(t) = p(0)\frac{\sin[\pi t/T]}{\pi t/T} \qquad [6.126]$$

This first Nyquist frequency criterion is only of theoretical interest. It has drawbacks which make it unusable in practice, since:

– the perfectly rectangular template $N_m(f)$ is impractical due to the gain discontinuity at the frequency $1/2T$;

– function $p(t)$ decreases as $1/|t|$, and a small error εT of the sampling instant $t_k = kT$ leads to a maximum $|I_{Max}|$ of ISI that is unbounded. Indeed, $|I_{Max}|$ in this case is given, according to [6.25], by:

$$|I_{Max}| = (M-1)\sum_{m=n-k,m\neq 0}\frac{|\sin[\varepsilon + m]|}{|\varepsilon + m|} \qquad [6.127]$$

The arithmetic series of general term $1/|\varepsilon + m|$ being divergent, so $|I_{Max}|$ is not bounded, that means that the eye pattern is completely closed at time $kT + \varepsilon T$.

Note that taking into account t_0, the first Nyquist criterion is expressed in the frequency domain and in the time domain respectively by:

$$P(f) = N_m(f)\exp(-j2\pi f t_0) \qquad [6.128]$$

and by:

$$p(t) = p(0)\frac{\sin[\pi(t-t_0)/T]}{\pi(t-t_0)/T} \qquad [6.129]$$

Third case

$$\frac{1}{T} = D_s < 2B$$

The folded spectrum consists of overlapping replicates of $P(f)$ (see Figure 6.12) and it is possible to find many functions $P(f)$ which satisfy the Nyquist frequency criterion called the "*extended Nyquist frequency criterion*" which do not have the drawbacks of the previous solution.

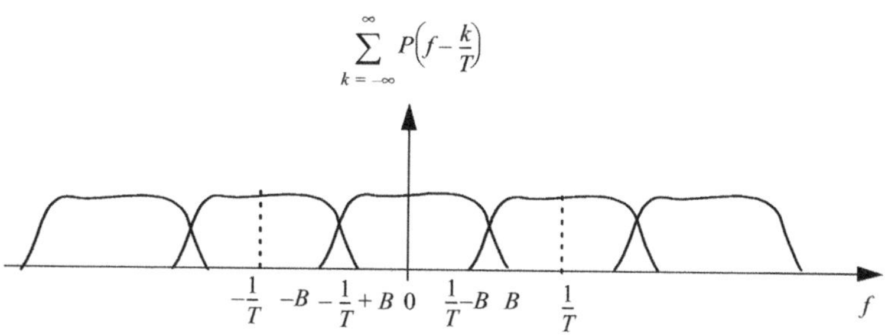

Figure 6.12. *Spectrum of* $\sum_{k=-\infty}^{\infty} P\left(f-\frac{k}{T}\right)$ *in the case where* $D_s < 2B$

In the case where $P(f) = 0$ for $|f| > 1/T$, real and positive *raised cosine functions* are widely used in practice in the transmitter and receiver filters of digital communications systems:

$$P_\alpha(f) = \begin{cases} T \times p(0) & \text{for} \quad 0 \leq |f| \leq \dfrac{1-\alpha}{2T} \\ \dfrac{T \times p(0)}{2}\left\{1 + \cos\left[\dfrac{\pi T}{\alpha}\left(|f| - \dfrac{1-\alpha}{2T}\right)\right]\right\} & \text{for} \quad \dfrac{1-\alpha}{2T} \leq |f| \leq \dfrac{1+\alpha}{2T} = f_{Max} \\ 0 & \text{for} \quad |f| > f_{Max} \end{cases} \qquad [6.130]$$

Parameter α, between 0 and 1, is called the "*roll-off factor*". Filters $P_\alpha(f)$ spread over the frequency bandwidth $\frac{1+\alpha}{2T}$ (by reasoning on the positive frequency domain), that is, an increase of the minimum bandwidth by a factor $1-\alpha$. The impulse response of these filters is of the form:

$$p_\alpha(t) = p(0) \times \frac{\sin[\pi t/T]}{\pi t/T} \times \frac{\cos[\alpha \pi t/T]}{1 - [2\alpha t/T]^2} \qquad [6.131]$$

Figures 6.13 and 6.14 show the normalized and delayed impulse response $p_\alpha(t)$ by $5T$ (so that the filter is causal) and the frequency response $P_\alpha(f)$ for different values of parameter α.

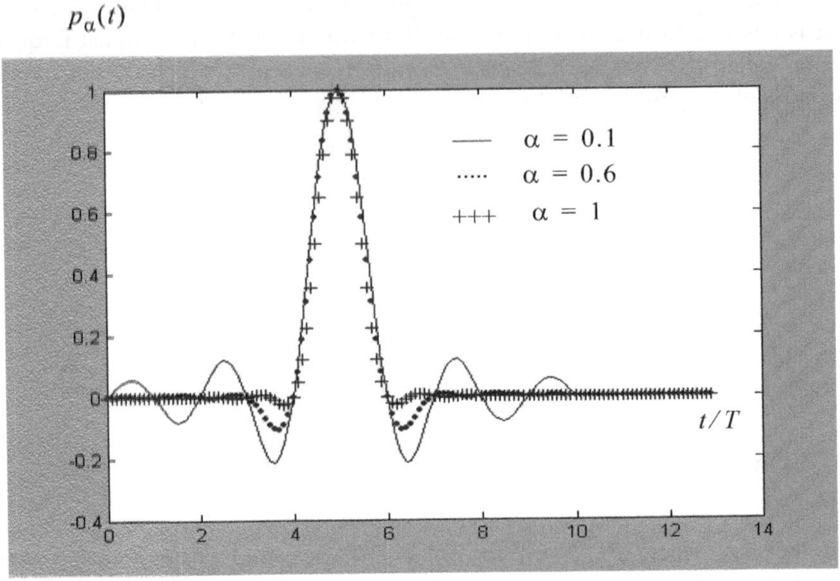

Figure 6.13. *Impulse response for different values of the parameter α. For a color version of this figure, see www.iste.co.uk/assad/digital1.zip*

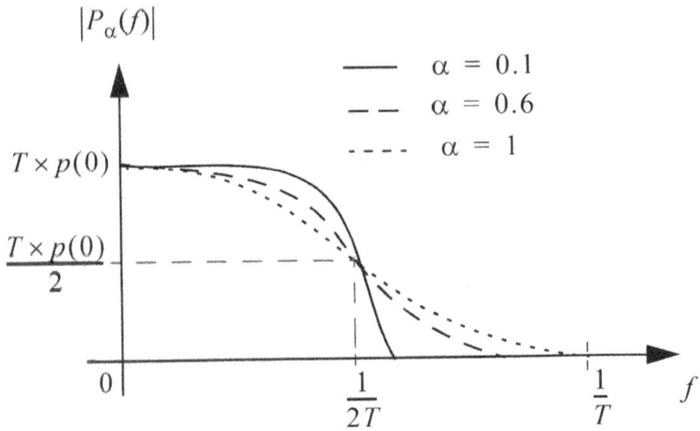

Figure 6.14. *Modulus of the frequency response of a raised cosine filter for different values of the parameter α*

For $\alpha = 0$, one gets the rectangular filter $N_m(f)$.

It can be seen that the amplitude of the oscillations of the function $p_\alpha(t)$ is smaller as the factor α is large, and therefore the bandwidth of the filter $P_\alpha(f)$ is large. This has an important practical consequence on the sensitivity of the receiver to inaccuracies in the sampling instants. Indeed, an offset εT of the sampling instant will introduce an ISI which is all the more significant as the amplitude of the oscillations is large. At the limit, when $\alpha = 1$, the amplitude of the oscillations is minimal (decay in $1/t^2$ of $p_\alpha(t)$), to the point that the function $p_\alpha(t)$ is canceled at times $(kT + T/2)$, for $k \neq [0, -1]$. The horizontal opening of the eye e_H is double compared to the case where $\alpha = 0$. This improvement comes at the expense of the necessary bandwidth of the channel, twice as much as that of Nyquist ideal.

In Figures 6.15, 6.16, 6.17, and 6.18, we represent the eye pattern in the case of binary symbols $a_n = \pm 1$ for several values of the parameter α. We notice that all the traces of the eye pattern pass through the same points at the sampling times in the absence of noise. We also show, in Figure 6.19, the eye pattern in the case of a quaternary symbol for $\alpha = 0.6$. Finally, note that the Nyquist filters are not attainable because the function $p_\alpha(t)$ is non-causal, $p_\alpha(t) \neq 0$ for $t < 0$, regardless of the value of the parameter α. However, in practice, the function $p_\alpha(t)$ is limited to a support fixed *a priori* with which we associate a delay line with a half-value delay in order to make causal the function $p_\alpha(t)$ as soon as $\alpha \geq 0.1$.

These filters are commonly used in digital communications systems with values of α in the order of 0.2 to 0.6.

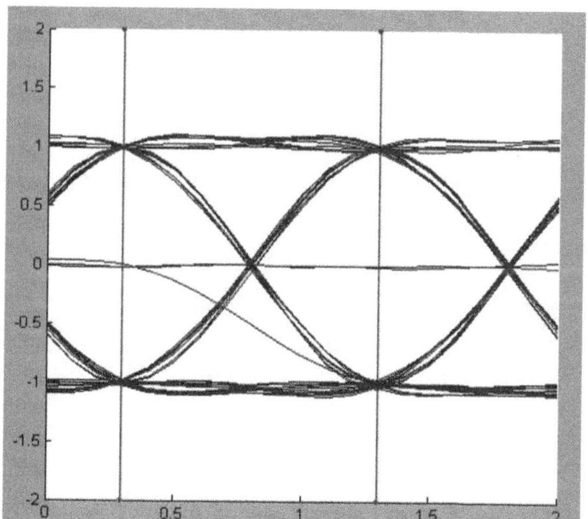

Figure 6.15. *Eye pattern (without noise), for α = 1;T = 1 s. For a color version of this figure, see www.iste.co.uk/assad/digital1.zip*

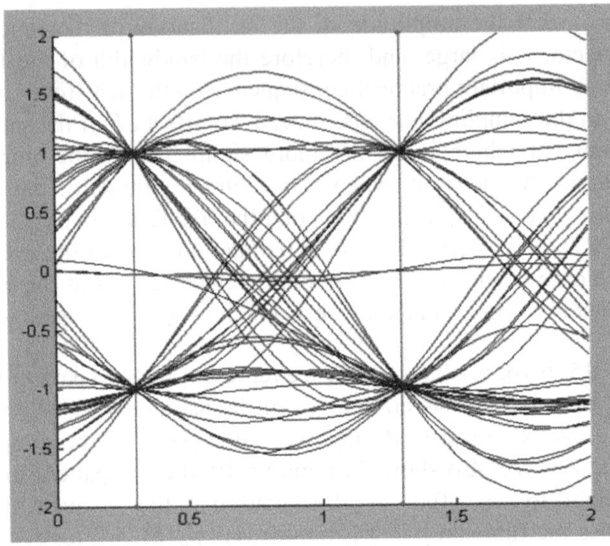

Figure 6.16. *Eye pattern (without noise), for α = 0.1; T = 1 s. For a color version of this figure, see www.iste.co.uk/assad/digital1.zip*

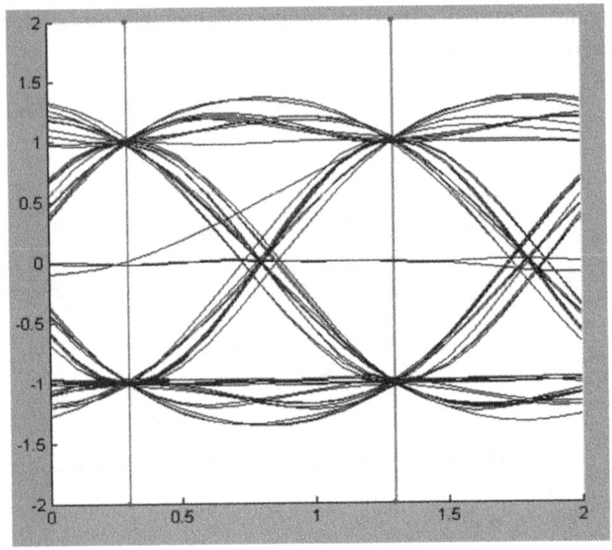

Figure 6.17. *Eye pattern (without noise), for α = 0.6; T = 1 s. For a color version of this figure, see www.iste.co.uk/assad/digital1.zip*

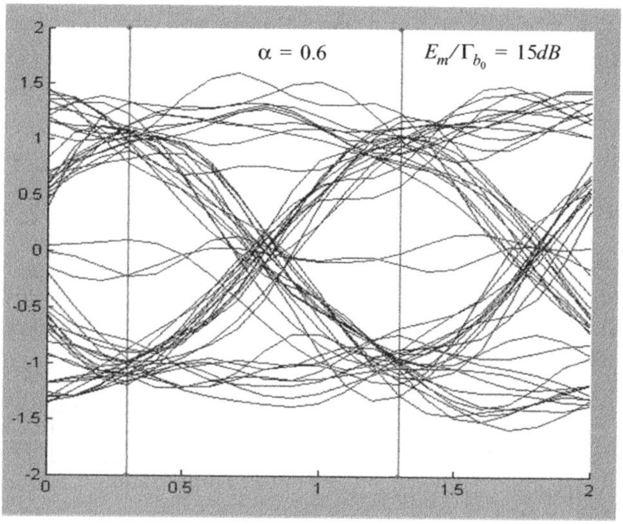

Figure 6.18. *Eye pattern (with noise), for $\alpha = 0.6$; $E_m/\Gamma_{b_0} = 15$ dB; $T = 1$ s. For a color version of this figure, see www.iste.co.uk/assad/digital1.zip*

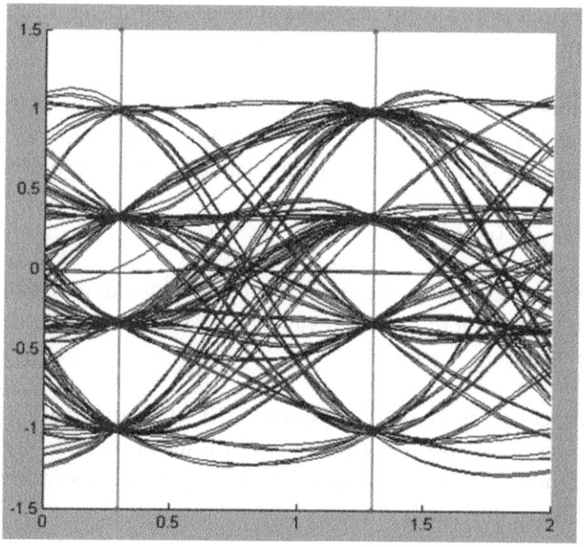

Figure 6.19. *Eye pattern 4-ary (without noise), for $\alpha = 0.6$; $T = 1$ s. For a color version of this figure, see www.iste.co.uk/assad/digital1.zip*

NOTE.– The Nyquist frequency criterion also applies when the function $P(f)$ is complex. Condition [6.120] becomes:

$$\begin{cases} \sum_k \Re[P(f-k/T)] = T \times p(0) \\ \sum_k \Im[P(f-k/T)] = 0 \end{cases} \qquad [6.132]$$

Furthermore, in the following, we show that the raised cosine filter which cancels the ISI is distributed between the transmitter and receiver filters. Each impulse response is then given by:

$$\sqrt{p_\alpha(t)} = \frac{4\alpha}{\pi\sqrt{T}} \times \frac{\left[\cos[(1+\alpha)\pi t/T] + \frac{\sin[(1-\alpha)\pi t/T]}{4\alpha t/T}\right]}{[1-(4\alpha t/T)^2]} \qquad [6.133]$$

6.7. Optimal distribution of filtering between transmission and reception

The problem dealt with in this section is that of optimizing all the elements of a transmission chain when it is affected by noise.

The cancellation of the ISI at the sampling times requires the equalization performed by the filter $G_c(f)$ to be such that the global transfer function of the transmission chain, see [6.12], satisfies the expanded Nyquist criterion:

$$X(f) \times H(f) \times G_c(f) = P_\alpha(f)\exp[-j2\pi f t_0] \qquad [6.134]$$

where, remember, t_0 represents the delay necessary for the emission and reception filters to be causal, and therefore achievable. It also includes the delay brought by the transmission channel.

In fact, in practice, as shown in Figure 6.20, the equalization filter $G_c(f)$ is distributed between the transmitter (transmission filter or preequalization $G_e(f)$) and the receiver (reception filter or post-equalization $G_r(f)$). This latter being is called the "equalizer". Thus, one has:

$$G_c(f) = G_e(f) \times G_r(f) \qquad [6.135]$$

This distribution makes it possible, among other things, to limit the bandwidth of the signal transmitted so as not to encroach on the adjacent channels. The reception filter limits the spectrum frequency of the received signal and thus eliminates part of the interference due to other channels and reduces the noise power.

It is now a question of finding the optimal filters of emission $G_e(f)$ and reception $G_r(f)$ which maximize the signal-to-noise ratio at the output of the reception filter (that is to say, at the input of the decision-making unit) at the sampling times. The noise $b_0(t)$ is considered here as a random, centered, 2nd order stationary Gaussian white process with a constant power spectral density $\Gamma_{b_0}(f)$, equal to Γ_0 in its frequency band and zero elsewhere. The noise $b_1(t)$ at the output of the reception filter is expressed by:

$$b_1(t) = b_0(t) \otimes g_r(t) \quad [6.136]$$

It has the same properties as noise $b_0(t)$. It power spectral density $\Gamma_{b_1}(f)$ and power $\sigma_{b_1}^2$ are given, respectively, by:

$$\Gamma_{b_1}(f) = \Gamma_{b_0}(f) \times |G_r(f)|^2 \quad [6.137]$$

and by:

$$P_{b_1}(f) = \sigma_{b_1}^2 = \int_{-\infty}^{\infty} \Gamma_{b_1}(f)df = \int_{-\infty}^{\infty} \Gamma_{b_0}(f) \times |G_r(f)|^2 df \quad [6.138]$$

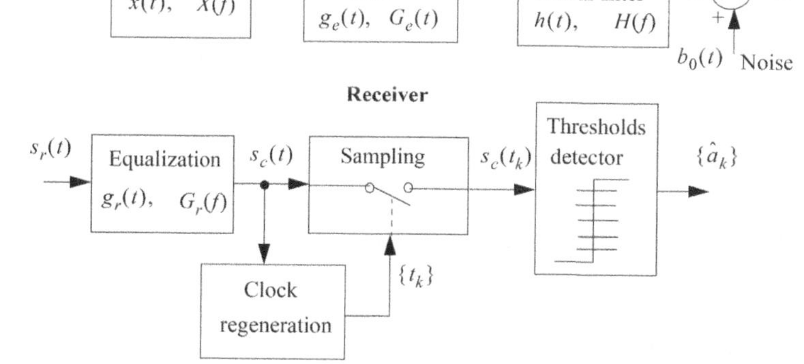

Figure 6.20. *Distribution of equalization filtering between transmitter and receiver*

In the absence of ISI, the output of the receiver filter at the sampling times $t_0 + kT$ is:

$$s_c(t_0 + kT) = a_k p(t_0) + b_1(t_0 + kT)] \quad [6.139]$$

The signal $s_c(t_0 + kT)$ is analogous to that obtained from an infinite bandwidth channel and the M-ary symbols are assumed to be equiprobable, we can show that, under these conditions, the probability of error P_e is given by:

$$P_e = \frac{M-1}{M} \operatorname{erfc}\left[\frac{p(t_0)}{\sigma_{b_1}\sqrt{2}}\right] \quad [6.140]$$

To minimize the probability of error, we have to maximize the signal-to-noise ratio in terms of power:

$$\left[\frac{S}{b}\right]_c^2 = \left[\frac{p(t_0)}{\sigma_{b_1}}\right]^2 = \left[\frac{V}{\sigma_{b_1}}\right]^2 \quad [6.141]$$

The amplitude V of the signal at the output of the M-ary to signal encoder is connected to the given average power P_m of the signal sent on the transmission medium.

One has:

$$s(t) = a(t) \otimes x(t) = \sum_n a_n\, x(t - nT) \quad [6.142]$$

and:

$$s_e(t) = s(t) \otimes g_e(t) \quad [6.143]$$

The average power P_m is written:

$$P_m = E\{|s_e(t)|^2\} = \frac{1}{T}\int_{-\infty}^{\infty} |s(t)|^2 \otimes |g_e(t)|^2 dt = \frac{V^2}{T}\int_{-\infty}^{\infty} |g_e(t)|^2 dt \quad [6.144]$$

But, also, according to Parseval's theorem:

$$P_m = \frac{V^2}{T}\int_{-\infty}^{\infty} |G_e(f)|^2 df \quad [6.145]$$

hence:

$$\frac{1}{V^2} = \frac{1}{P_m T} \int_{-\infty}^{\infty} |G_e(f)|^2 df \qquad [6.146]$$

The filter $G_e(f)$ should fulfill the Nyquist criterion given by [6.134]:

$$|G_e(f)| = \begin{cases} \dfrac{|P_\alpha(f)|}{|X(f)||H(f)||G_r(f)|} & \text{for } |f| \leq B \\ 0 & \text{elsewhere} \end{cases} \qquad [6.147]$$

hence:

$$\frac{1}{V^2} = \frac{1}{P_m T} \int_{-B}^{B} \frac{|P_\alpha(f)|^2}{|X(f)|^2 |H(f)|^2 |G_r(f)|^2} df \qquad [6.148]$$

Maximizing the signal-to-noise ratio means minimizing the noise-to-signal ratio:

$$\left[\frac{\sigma_{b_1}}{V}\right]^2 = \frac{1}{P_m T} \int_{-B}^{B} \Gamma_{b_0}(f) |G_r(f)|^2 df \int_{-B}^{B} \frac{|P_\alpha(f)|^2}{|X(f)|^2 |H(f)|^2 |G_r(f)|^2} df \qquad [6.149]$$

The Cauchy–Schwartz inequality is written:

$$\int_{-\infty}^{\infty} |U_1(f)|^2 \, df \times \int_{-\infty}^{\infty} |U_2(f)|^2 \, df \geq \left[\int_{-\infty}^{\infty} |U_1(f)| \times |U_2(f)| df \right]^2 \qquad [6.150]$$

Let:

$$\begin{cases} |U_1(f)| = [\Gamma_{b_0}(f)]^{1/2} |G_r(f)| \\ |U_2(f)| = \dfrac{|P_\alpha(f)|}{|X(f)||H(f)||G_r(f)|} \end{cases} \qquad [6.151]$$

Then the modulus of the reception filter $|G_r(f)|$ (post-equalization filter) that minimizes the ratio $\left[\frac{\sigma_{b_1}}{V}\right]^2$ is given for:

$$U_1(f) = K_1 U_2(f) \qquad [6.152]$$

where K_1 is any real proportionality coefficient.

So:

$$|G_r(f)| = K \times \frac{|P_\alpha(f)|^{1/2}}{[\Gamma_{b_0}(f)]^{1/4}|X(f)|^{1/2}|H(f)|^{1/2}} \quad [6.153]$$

for $|f| \leq B$

with $K = \sqrt{K_1}$.

From [6.147] the modulus of the optimal transmitter filter is expressed by:

$$|G_e(f)| = \frac{1}{K} \times \frac{|P_\alpha(f)|^{1/2}[\Gamma_{b_0}(f)]^{1/4}}{|X(f)|^{1/2}|H(f)|^{1/2}} \quad [6.154]$$

In these conditions, the maximum of the signal-to-noise ratio is:

$$\left[\frac{S}{b}\right]^2_{c,Max} = \left[\frac{V}{\sigma_{b_1}}\right]^2_{Max} = \frac{P_m T}{\left[\int_{-B}^{B}[\Gamma_{b_0}(f)]^{1/2}\frac{|P_\alpha(f)|}{|X(f)||H(f)|}df\right]^2}$$

$$= \frac{P_m T}{\left[\int_{-B}^{B}[\Gamma_{b_0}(f)]^{1/2}|G_e(f)||G_r(f)|df\right]^2} \quad [6.155]$$

The optimal transmitter and receiver filters are only defined in modules. The phases of the different filters are chosen so as to satisfy the Nyquist criterion [6.134], hence:

$$\phi_x + \phi_e + \phi_h + \phi_r = 2\pi f t_0 \quad [6.156]$$

PRACTICAL NOTE.– The preequalization in the transmitter by the optimal emission filter given by relation [6.154] raises practical problems of realization and compatibility between materials of different origins. For this, the principle generally adopted in standardization consists of imposing the characteristics of the signal applied to the transmission medium and leaving maximum freedom for the realization of compatible receivers. For this reason, we define the Nyquist filter $P_\alpha(f)$ and impose the spectrum of the transmitted signal:

$$|G_e(f)| = \frac{1}{K} \times \frac{|P_\alpha(f)|^{1/2}[\Gamma_{b_0}(f)]^{1/4}}{|X(f)|} \quad \text{for } |f| \leq B \quad [6.157]$$

The reception filter is then adjusted according to the characteristics $|H(f)|$ of the transmission medium:

$$|G_r(f)| = K \times \frac{|P_\alpha(f)|^{1/2}}{\left[\Gamma_{b_0}(f)\right]^{1/4}|H(f)|} \quad \text{for} \quad |f| \leq B \qquad [6.158]$$

Filters $|G_e(f)|$ and $|G_r(f)|$, thus defined, allow the cancellation of the ISI, but not the maximization of the signal-to-noise ratio at the output of the receiver.

A more efficient but more complex method consists of optimizing the signal-to-noise ratio by applying the condition [6.141] and taking account of the ISI in the estimation process, by carrying out the decision-making using several consecutive samples. An estimate of the sequence of symbols emitted is thus made according to the *maximum likelihood a posteriori*, that is to say, by determining and maximizing the probability *a posteriori* that the estimated sequence has been transmitted, knowing the sequence of samples received. This process requires adaptation of the receiver to the characteristics of the transmission channel.

6.7.1. *Expression of the minimum probability of error for a low-pass channel satisfying the Nyquist criterion*

The probability of error of M-ary symbols a_k is given by:

$$P_e = \frac{M-1}{M} \operatorname{erfc}\left[\frac{p(t_0)}{\sigma_{b_1}\sqrt{2}}\right]$$

This probability of error P_e is minimal when $\left[\frac{p(t_0)}{\sigma_{b_1}}\right]$ is maximal. But, according to the expression [6.155], we recall:

$$\left[\frac{p(t_0)}{\sigma_{b_1}}\right]^2 = \frac{P_m T}{\int_{-B}^{B}\left[\left[\Gamma_{b_0}(f)\right]^{1/2}|G_e(f)||G_r(f)|df\right]^2}$$

The ratio is maximal if the reception filter is matched to the emission filter:

$$g_r(t) = k \times g_e(t-\tau) \qquad [6.159]$$

Moreover, if the noise $b_0(t)$ is white noise, then:

$$\left[\frac{p(t_0)}{\sigma_{b_1}}\right]^2_{Max} = \left[\frac{2P_mT}{\Gamma_{b_0}(f)}\right] = \left[\frac{2E_a}{\Gamma_{b_0}(f)}\right] \qquad [6.160]$$

where E_a is the average energy required to transmit a symbol a_k.

The probability of error is finally expressed by:

$$P_e = \frac{M-1}{M}\operatorname{erfc}\left[\sqrt{\frac{E_a}{\Gamma_{b_0}(f)}}\right] \qquad [6.161]$$

6.8. Transmission with a partial response linear coder

Partial response linear coding introduces, in a controlled manner, an interference between symbols, known to the receiver and therefore able to be taken into account during decoding. The advantage of such coding is that by accepting a certain amount of ISI, it is possible to transmit a signal with a symbol rate $D_s = 1/T$ on a channel having a bandwidth strictly equal to $1/2T$.

Two partial response codes characterized by their transfer function are used here to illustrate the result cited above.

The two codes analyzed and evaluated subsequently are:

– the duobinary code, characterized by its transfer function:

$$H(z) = 1 + z^{-1} \qquad [6.162]$$

– the interleaved bipolar code of order 2 whose transfer function is given by:

$$H(z) = 1 - z^{-2} \qquad [6.163]$$

Figure 6.21 shows the transmission and reception chain with a partial response linear coding. We then adopt the same approach as A. Glavieux and M. Joindot in (Glavieux and Joindot 1996).

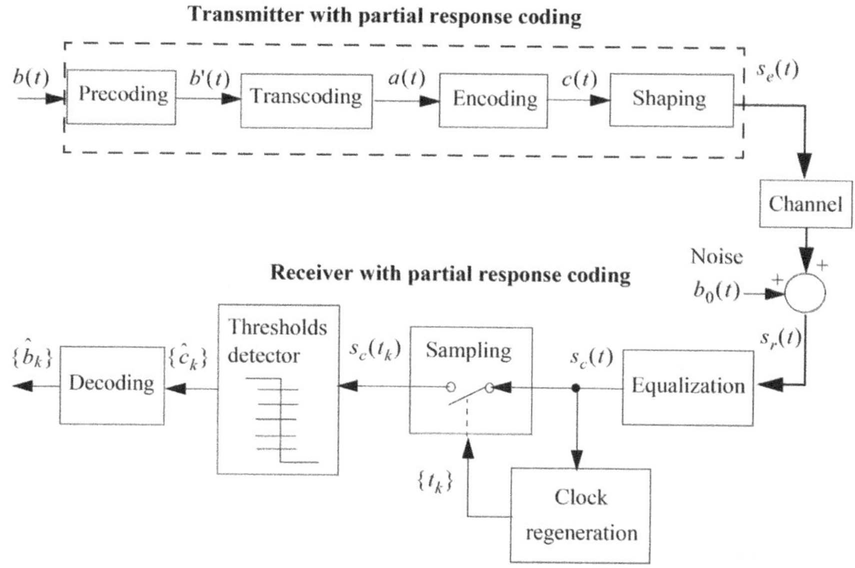

Figure 6.21. *Transmission and reception chain with partial response linear coding*

6.8.1. *Transmission using the duobinary code*

Let us consider the transmission of a series of binary symbols a_n over a noisy low-pass channel with bandwidth B. The sample $s_c(t_0 + kT)$ at the input of the decision block is written in the usual form:

$$s_c(t_0 + kT) = a_k p(t_0) + \sum_{n \neq k} a_n p[t_0 + (k-n)T] + b_1(t_0 + kT) \qquad [6.164]$$

Let us consider that only the symbol a_{k-1} interferes with the symbol a_k, thus producing an ISI equal to $a_{k-1} p(t_0 + T)$.

Under these conditions, and assuming that:

$$p(t_0 + T) = p(t_0) \qquad [6.165]$$

The expression of $s_c(t_0 + kT)$ is then:

$$s_c(t_0 + kT) = [a_k + a_{k-1}]\, p(t_0) + b_1(t_0 + kT) \qquad [6.166]$$

The temporal condition of this particular form of ISI is:

$$p(t_0 + kT) = \begin{cases} p(t_0) & \text{for } k = 0, 1 \\ 0 & \text{elsewhere} \end{cases} \qquad [6.167]$$

The frequency function $P(f)$ that leads to this same ISI should satisfy the following equality:

$$\sum_{k=-\infty}^{\infty} P\left(f - \frac{k}{T}\right) \exp\left[j2\pi\left(f - \frac{k}{T}\right)t_0\right] = T \times p(t_0)[1 + \exp(-j2\pi fT)] \qquad [6.168]$$

Or again:

$$\sum_{k=-\infty}^{\infty} P\left(f - \frac{k}{T}\right) \exp\left[j2\pi\left(f - \frac{k}{T}\right)t_0\right] = 2T \times p(t_0)\cos(\pi fT)\exp(-j\pi fT) \qquad [6.169]$$

The condition [6.169] is fulfilled by filter $P(f)$:

$$P(f) = |P(f)|\exp[j\phi(f)] \qquad [6.170]$$

that satisfies:

$$|P(f)| = \begin{cases} 2T \times p(t_0)|\cos(\pi fT)| & \text{for } |f| \le 1/2T \\ 0 & \text{elsewhere} \end{cases}$$

$$\phi(f) = -2\pi f(t_0 + T/2) \qquad [6.171]$$

This filter covers a frequency band of $1/2T$ (by reasoning on the positive frequencies) but, unlike the filter verifying the first Nyquist criterion, it does not present discontinuities at frequencies $\pm 1/2T$ and it is therefore physically achievable (see Figure 6.22, section 6.8.3). Furthermore, the maximum distortion due to an offset ε of the optimal sampling instant $(t_0 + kT)$ remains finite. The eye pattern is therefore opened in the vicinity of the optimal decision instant. Furthermore, by setting $c_k = a_k + a_{k-1}$, the expression [6.166] is written:

$$s_c(t_0 + kT) = c_k\, p(t_0) + b_1(t_0 + kT) \qquad [6.172]$$

NOTE.– The expression [6.172] shows that the transmission of a sequence of independent a_n binary symbols, in the presence of a particular ISI, is seen as the transmission of a sequence of correlated ternary symbols c_n, but without ISI.

6.8.1.1. *Demonstration of relation [6.168]*

To demonstrate the relationship, we adopt the approach explained in section 6.6.

The sampled signal $p_\Delta(t)$ is written:

$$p_\Delta(t) = p(t) \times \sum_{i=-\infty}^{\infty} \delta(t - iT - t_0) = \sum_{i=-\infty}^{\infty} p(iT + t_0)\delta(t - iT - t_0) \qquad [6.173]$$

Or considering relations [6.166] and [6.167]:

$$p_\Delta(t) = p(t_0)[\delta(t - t_0) + \delta(t - T - t_0)] \qquad [6.174]$$

The respective Fourier transforms of expressions [6.173] and [6.174] are written:

$$P_\Delta(f) = P(f) \otimes \frac{1}{T} \sum_{k=-\infty}^{\infty} \delta\left(f - \frac{k}{T}\right) \exp\left(-j2\pi \frac{k}{T} t_0\right) =$$

$$\frac{1}{T} \sum_{k=-\infty}^{\infty} P\left(f - \frac{k}{T}\right) \exp\left(-j2\pi \frac{k}{T} t_0\right) \qquad [6.175]$$

$$P_\Delta(f) = \int_{-\infty}^{\infty} p(t_0)[\delta(t - t_0) + \delta(t - T - t_0)]\exp(-j2\pi f t)\, dt$$

$$= p(t_0)\big[\exp(-j2\pi f t_0) + \exp\big(-j2\pi f(T + t_0)\big)\big]$$

$$= p(t_0)\exp(-j2\pi f t_0)[1 + \exp(-j2\pi f T)] \qquad [6.176]$$

By equalizing the two expressions of $P_\Delta(f)$, we obtain relation [6.168].

6.8.2. *Transmission using 2nd order interleaved bipolar code*

In this case, the ISI added to the useful signal at the instant $(t_0 + kT)$ is:

$$-a_{k-2}p(t_0 + 2T)$$

The sample $s_c(t_0 + kT)$ at the output of the reception filter is equal to:

$$s_c(t_0 + kT) = a_k\, p(t_0) - a_{k-2}\, p(t_0 + 2T) + b_1(t_0 + kT) \qquad [6.177]$$

again, assuming that $p(t_0 + 2T) = p(t_0)$:

$$s_c(t_0 + kT) = [a_k - a_{k-2}]p(t_0) + b_1(t_0 + kT) \qquad [6.178]$$

To obtain this amount of ISI, the signal $p(t)$ must fulfill the following temporal condition:

$$p(t_0 + iT) = \begin{cases} p(t_0) & \text{for } i = 0, 2 \\ 0 & \text{elsewhere} \end{cases} \qquad [6.179]$$

By following the same approach as before, we easily show that the function $P(f)$, which leads to the ISI mentioned above, verifies the following relation:

$$\sum_{k=-\infty}^{\infty} P\left(f - \frac{k}{T}\right) \exp\left[j2\pi\left(f - \frac{k}{T}\right) t_0\right] = T \times p(t_0)[1 - \exp(-j4\pi fT)] \qquad [6.180]$$

which is fulfilled by a filter $P(f)$ such that:

$$\begin{aligned} |P(f)| &= \begin{cases} 2T \times p(t_0)|\sin(2\pi fT)| & \text{for } |f| \leq 1/2T \\ 0 & \text{elsewhere} \end{cases} \\ \phi(f) &= -2\pi f[t_0 + T] + \pi/2 \end{aligned} \qquad [6.181]$$

The modulus $|P(f)|$ of the frequency response is shown in Figure 6.22.

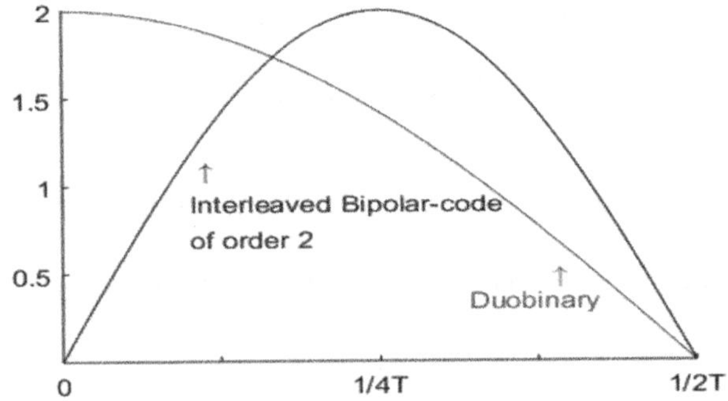

Figure 6.22. *Modulus normalized with respect to $T \times p(t_0)$ of the frequency response $|P(f)|$ of codes duobinary and bipolar interleaved of order 2. For a color version of this figure, see www.iste.co.uk/assad/digital1.zip*

As in the case of the duobinary code, the frequency response occupies a frequency band equal to $1/2T$ (for positive frequencies) and has no discontinuity. In addition, the interleaved bipolar code of order 2 can be used for transmission over a cable, since $|P(f)|$ at $f = 0$ is zero.

By setting $c_k = a_k - a_{k-2}$, the expression [6.178] is written like the expression [6.172]:

$$s_c(t_0 + kT) = c_k\, p(t_0) + b_1(t_0 + kT) \qquad [6.182]$$

and the remark from the previous paragraph concerning the duobinary code applies.

6.8.3. *Reception of partial response linear codes*

In the previous section, it was shown that the transmission of binary information in the presence of controlled ISI is transformed, thanks to partial response coding, to the transmission of a series of correlated M-ary symbols without ISI.

Recall the relationship between the symbols c_k and a_k which depends on the type of ISI considered here, namely:

$$\begin{cases} c_k = a_k + a_{k-1} & \text{if } H(z) = 1 + z^{-1} \\ c_k = a_k - a_{k-2} & \text{if } H(z) = 1 - z^{-2} \end{cases} \qquad [6.183]$$

Decoding of binary symbols $\hat{b}_k$ is obtained directly from the symbols $\hat{c}_k$ without propagation of error, thanks to the function of precoding on transmission.

6.8.3.1. *Case of duobinary coding*

The precoder generates a sequence $\{b'_k\}$ such as:

$$b'_k = b_k \oplus b'_{k-1} \qquad [6.184]$$

The transcoder generates, at output, a symmetrical binary sequence $\{a_k\}$:

$$a_k = 2b'_k - 1 \qquad [6.185]$$

such that it generates:

$$\begin{cases} b'_k = 0 & \rightarrow a_k = -1 \\ b'_k = 1 & \rightarrow a_k = 1 \end{cases} \qquad [6.186]$$

The relation defining the operation of the partial response coder is expressed by:

$$c_k = a_k + a_{k-1} = [2b'_k - 1 + 2b'_{k-1} - 1] = 2[b'_k + b'_{k-1} - 1] \qquad [6.187]$$

with $c_k \in [0, \pm 2]$.

Then, according to [6.187], we can write:

$$\frac{1}{2}c_k + 1 = b'_k + b'_{k-1} = b_k$$

then:

$$\hat{b}_k = \frac{1}{2}\hat{c}_k + 1 \mod 2 \qquad [6.188]$$

So a direct estimate of the transmitted binary sequence $\{\hat{b}_k\}$ from the estimated sequence of symbols $\{\hat{c}_k\}$ received is:

$$\begin{cases} \hat{c}_k = 0 \rightarrow \hat{b}_k = 1 \\ \hat{c}_k = \pm 2 \rightarrow \hat{b}_k = 0 \end{cases} \qquad [6.189]$$

6.8.3.2. *Example of duobinary coding and decoding*

In Figure 6.23, we give an example of duobinary coding and decoding.

$\{b_k\}$		0	0	1	0	1	1	0
$\{b'_k\}$	**0**	0	0	1	1	0	1	1
$\{a_k\}$	**-1**	-1	-1	1	1	-1	1	1
$\{\hat{c}_k\}$		-2	-2	0	2	0	0	2
$\{\hat{b}_k\}$		0	0	1	0	1	1	0

Figure 6.23. *Example of duobinary coding and decoding*

We can easily see that the absence of precoding on transmission would lead to a propagation of decision errors on the binary symbols $\{b_k\}$. Indeed, in this case, according to [6.187], by replacing b'_k by b_k, one has:

$$\hat{b}_k = \frac{1}{2}\hat{c}_k + 1 - \hat{b}_{k-1} \mod 2 \quad [6.190]$$

So, if the binary element b_{k-1} is badly decoded, the binary element b_k will also almost certainly be.

6.8.3.3. *Case of 2nd order interleaved bipolar coding*

In this case, the precoder is defined by the following rule:

$$b'_k = b_k \oplus b'_{k-2} \quad [6.191]$$

The transcoding operation is the same whatever the code used, and it is therefore governed by relation [6.185].

The relation defining the operation of the partial response coder is written:

$$c_k = a_k - a_{k-2} = [2b'_k - 1 - (2b'_{k-2} - 1)] = 2[b'_k - b'_{k-2}] = 2b_k \quad [6.192]$$

with $c_k \in [0, \pm 2]$.

Hence:

$$\hat{b}_k = \frac{1}{2}\hat{c}_k \mod 2 \quad [6.193]$$

This relation shows the instantaneous estimate of the sequence $\{\hat{b}_k\}$ sent from the sequence $\{\hat{c}_k\}$ received:

$$\begin{cases} \hat{c}_k = 0 \rightarrow \hat{b}_k = 0 \\ \hat{c}_k = \pm 2 \rightarrow \hat{b}_k = 1 \end{cases} \quad [6.194]$$

EXAMPLE.– 2nd order interleaved bipolar coding and decoding.

Figure 6.24 gives an example of 2nd order interleaved bipolar coding and decoding.

$\{b_k\}$		0	0	1	0	1	1	0
$\{b'_k\}$	**0 0**	0	0	1	0	0	1	0
$\{a_k\}$	**-1 -1**	-1	-1	1	-1	-1	1	-1
$\{\hat{c}_k\}$		0	0	2	0	-2	2	0
$\{\hat{b}_k\}$		0	0	1	0	1	1	0

Figure 6.24. *Example of 2nd order interleaved bipolar coding and decoding*

6.8.4. *Probability of error P_e*

Insofar as the binary elements b_k are independent and equiprobable on the alphabet $\{0, 1\}$, the symbols c_k take the values $\{-2, 0, 2\}$ with, respectively, the following probabilities: $\left\{\frac{1}{4}, \frac{1}{2}, \frac{1}{4}\right\}$.

The probability of error is then equal to:

$$P_e = \frac{1}{4} P_{e_{-2}} + \frac{1}{2} P_{e_0} + \frac{1}{4} P_{e_2} \qquad [6.195]$$

where P_{e_k} is the conditional probability that $c_k = k$ with $k = \{-2, 0, 2\}$.

Let:

$$P_{e_2} = Pr\{\hat{c}_k \neq 2/c_k = 2\}$$
$$= Pr\{\hat{c}_k = -2/c_k = 2\} + Pr\{\hat{c}_k = 0/c_k = 2\} \qquad [6.196]$$

$$P_{e_{-2}} = Pr\{\hat{c}_k \neq -2/c_k = -2\}$$
$$= Pr\{\hat{c}_k = 2/c_k = -2\} + Pr\{\hat{c}_k = 0/c_k = -2\} \qquad [6.197]$$

and:

$$P_{e_0} = Pr\{\hat{c}_k \neq 0/c_k = 0\}$$
$$= Pr\{\hat{c}_k = 2/c_k = 0\} + Pr\{\hat{c}_k = -2/c_k = 0\} \qquad [6.198]$$

By comparing the sample $s_c(t_0 + kT)$ to a pair of thresholds $-p(t_0)$ and $p(t_0)$, we can easily calculate the different expressions of conditional probabilities.

6.8.4.1. *Calculation of conditional probabilities*

This now classic calculation is thus developed (see also Figure 6.25):

$$b_1(t_0 + kT) = s_c(t_0 + kT) - c_k\, p(t_0) \qquad [6.199]$$

if:

$$s_c(t_0 + kT) \in [c, d[\ ; c, d \in \{\mu_m, \pm\infty\} \qquad [6.200]$$

where:

$$\mu_m = m \times p(t_0) \quad \text{with} \quad m = \{-1, 1\} \qquad [6.201]$$

so:

$$b_1(t_0 + kT) \in [c - c_k p(t_0), d - c_k p(t_0)[\qquad [6.202]$$

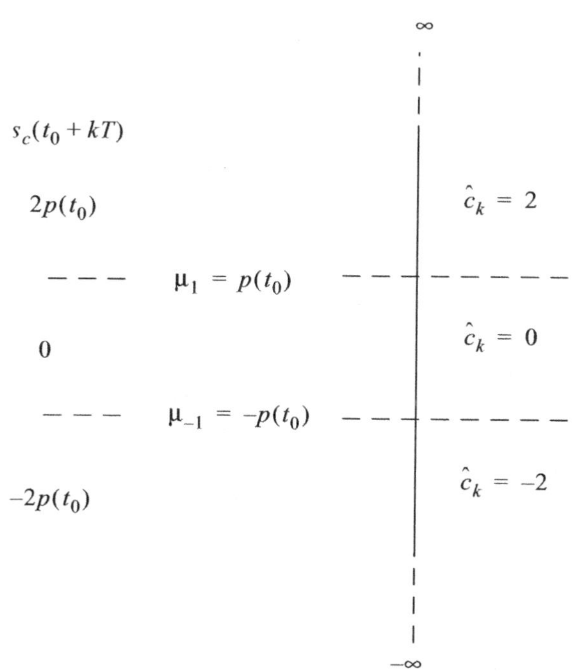

Figure 6.25. *Sample values, optimal thresholds and estimation classes* $c_k : \hat{c}_k$

Application of the decision rules

1) $\hat{c}_k = 0 \Leftrightarrow s_c(t_0 + kT) \in [-p(t_0), p(t_0)[$

$$-p(t_0) - c_k p(t_0) \leq b_1(t_0 + kT) < p(t_0) - c_k p(t_0) \quad [6.203]$$

hence:

$$Pr\{\hat{c}_k = 0/c_k = -2\} = Pr\{p(t_0) \leq b_1(t_0 + kT) < 3p(t_0)\} \quad [6.204]$$

$$Pr\{\hat{c}_k = 0/c_k = 2\} = Pr\{-3p(t_0) \leq b_1(t_0 + kT) < -p(t_0)\} \quad [6.205]$$

2) $\hat{c}_k = 2 \Leftrightarrow s_c(t_0 + kT) \in [p(t_0), \infty[$

$$p(t_0) - c_k p(t_0) \leq b_1(t_0 + kT) < \infty \quad [6.206]$$

hence:

$$Pr\{\hat{c}_k = 2/c_k = 0\} = Pr\{p(t_0) \leq b_1(t_0 + kT) < \infty\} \quad [6.207]$$

$$Pr\{\hat{c}_k = 2/c_k = -2\} = Pr\{3p(t_0) \leq b_1(t_0 + kT) < \infty\} \quad [6.208]$$

3) $\hat{c}_k = -2 \Leftrightarrow s_c(t_0 + kT) \in [-\infty, -p(t_0)[$

$$-\infty \leq b_1(t_0 + kT) < -p(t_0) - c_k p(t_0) \quad [6.209]$$

hence:

$$Pr\{\hat{c}_k = -2/c_k = 0\} = Pr\{-\infty \leq b_1(t_0 + kT) < -p(t_0)\} \quad [6.210]$$

$$Pr\{\hat{c}_k = -2/c_k = 2\} = Pr\{-\infty \leq b_1(t_0 + kT) < -3p(t_0)\} \quad [6.211]$$

NOTE.– As soon as the signal-to-noise ratio E_{mb}/Γ_{b_0} is sufficiently large, one can make the reasonable assumption of neglecting the possibility of decoding $\hat{c}_k = -2$ when $c_k = 2$ or vice versa. The probability of error on the binary digit P_{eb} is equal to the probability of error P_e on the symbols c.

7

Digital Transmissions with Carrier Modulation

7.1. Introduction and schematic diagram of a digital radio transmission

In Chapter 6, we studied the methods used for digital baseband transmissions where the transmission medium is likened to a low-pass filter. However, most of the transmission media are of the *bandpass type*. This is the case for:

– the analog telephone channel for which the authorized frequency band resulting from all the filters involved is 300 Hz–3,400 Hz;

– the radio channel (radio communications), the case of the propagation of terrestrial electromagnetic waves (EMW), for which the attenuation decreases when the frequency increases (law in $1/f^2$). The presence of the ionospheric layer causes refractions leading to non-linear propagation and the generation of multiple propagation paths. In addition, the absorption of waves due to water vapor in the troposphere causes significant attenuations at very high frequencies (several tens of GHz) but selectively (in certain frequency bands);

– the earth-satellite channel, the case of the propagation of EMWs in space, comparable to terrestrial hertzian transmissions but with the distinction that ionospheric refraction is much less important because of much more favorable angles of incidence.

In all these situations, the technique used consists of transmitting a signal whose frequencies are compatible with the transmission channel and obtained by modulating a single-frequency signal by the digital baseband signal. Moreover, in radio transmissions, for reasons related to the simultaneous transport of several channels in frequency bands adjacent to each other, the digital signal first modulates an intermediate carrier at a frequency f_I (for example, 70 MHz). The resulting signal, or the sum of the resulting signals in the case of a frequency multiplex, then modulates the carrier at the frequency f_C (a few GHz in the case of the UMTS standard – Universal Mobile Telecommunications System), producing the actual transmitted signal.

The set of the transmitter, the transmission channel and the receiver is represented in principle in Figure 7.1. We have not only included the noise resulting from the transmission channel but also the propagation interferences corresponding to the presence of multiple propagation paths, interference from the other transmitters working in the same frequency domains and, finally, non-linearities of the signal caused by the power amplifier. The letter S in Figure 7.1, denotes the frame synchronization.

In this chapter, we will detail the specificities of digital transmission with carrier modulation compared to a digital baseband transmission:

– *types* and *properties of digital modulations*;

– *types* and *characteristics of radio impairments* applying to the transmitted signal. We will specify the influence of different filtering on intersymbol interference and noises. In particular, it will be seen that the *equivalent baseband channel model* makes it possible to treat the problem, in the case of linear modulation, in a way that is identical to that of the digital baseband transmission systems. Only the amplitude of the signals to be considered is modified, it will be complex and no longer real.

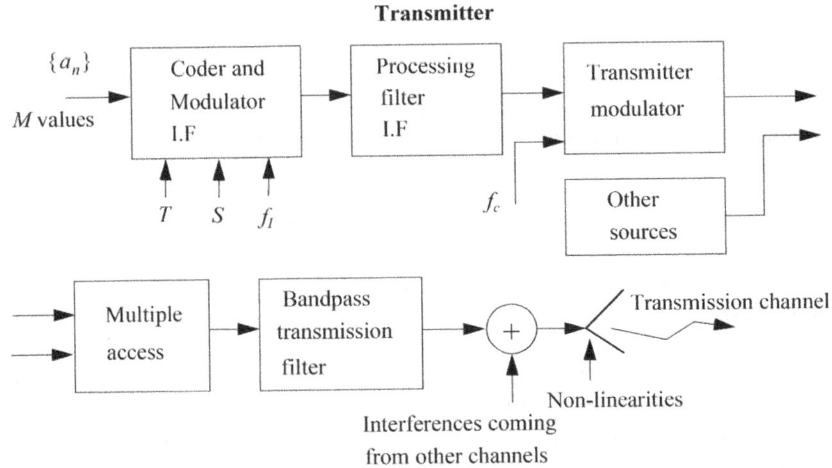

Figure 7.1(a). *Schematic block diagram of a digital radio transmission system*

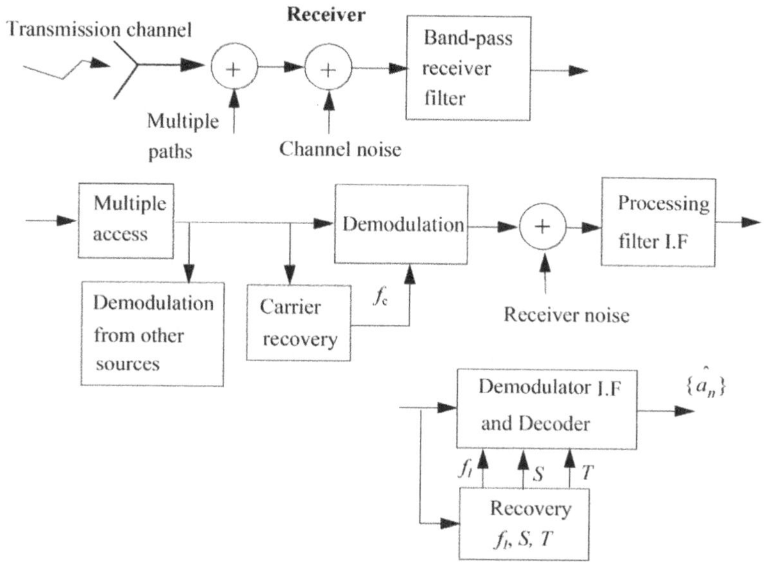

Figure 7.1(b). *Schematic block diagram of a digital radio transmission system (following)*

7.2. Multiple access techniques and most common standards

Several users can communicate at the same time on the same transmission channel, thanks to multiple access techniques. These techniques are grouped into two broad classes: the dynamic access technique used in the access control sublayer to the channel, known as Medium Access Control (MAC), and the traditional so-called static multiple access techniques used in the physical layer. We summarize hereinafter the static multiple access techniques:

– Frequency Division Multiple Access (FDMA) in which the available frequency band of the channel is divided into narrower bands (sub-channels), each of which is allocated for the exclusive use of a user (a subscriber);

– Time Division Multiple Access (TDMA) in which each user uses the entire available frequency band of the transmission channel but in more or less regularly spaced time slots;

– Direct Sequence Code Division Multiple Access (DS-CDMA), thanks to the spread spectrum technique, in which the multiplexing is done by both frequency division and time division. In this method, each user is assigned a unique code sequence (signature sequence) that allows him/her to spread the information-bearing signal accross the assigned frequency band. Thus, at the receiver, signals from the different users are separated by cross-correlating the received signal with each of the possible user signature sequences.

In Figure 7.2, we schematically show the principle of these three multiple access techniques and in Figure 7.3, we give the principle of direct sequence spread spectrum for a user.

The DS-CDMA technique is used in the UMTS: for the mobile uplink to the station, each user has a pseudo-random Gold type code with a chip rate $D_c = 3.84\,\text{Mc/s}$ and a symbol-rate $D_s = 30\,\text{Kb/s}$ that corresponds to the transmission of the voice after encoding and interleaving. Hence the maximum number of users who can communicate at the same time $Nb_{max} = D_c/D_s = 128$ users.

– Orthogonal Frequency Division Multiple Access (OFDM). In the multiple access technique, the spectrum is divided into multiple orthogonal subcarriers between them thanks to the Inverse Fast Fourier Transform (IFFT), so that at the frequency at which a carrier has its maximum peak of power, at this same frequency the other carriers have zero power and this is verified regardless of the carrier considered. This avoids interference between the subcarriers. Data is transmitted simultaneously on each phase modulated (PSK) or quadrature amplitude modulated (QAM) subcarrier. This technique allows very high data rates (up to 54 Mbit/s) and effectively withstands the effects of multiple paths. It is used in: WiFi (Wireless Fidelity), DSL (Digital Subscriber Line), DVB (Digital Audio Broadcasting) and 4G radio communication (LTE: Long Term Evolution).

Several variants of OFDM exist:

– *Discrete Multitone Modulation* (DMT) refers to an OFDM baseband transmission system;

– *Coded Orthogonal Frequency Division Multiplexing* (COFDM) introduces an error correction code (Reed Solomon and Viterbi);

– *Weighted Cyclic Prefix Orthogonal Frequency Division Multiplexing* (WCP-OFDM) provides for the addition of a cyclic prefix and the weighting of the output signal of the transmitter in order to adapt to multipath mobile channels;

– *Orthogonal Frequency Division Multiple Access* (OFDMA) is derived from *OFDM*. It can be described as a combination of OFDM with TDMA. It is used in 4G LTE mobile radio communication, and mobile WiMAX.

In Figure 7.4, we give the allocation in France of the OFDM spectrum in the 802.11.g standard.

In France, the 802.11.g standard uses 11 channels. The first has a central frequency of 2,412 MHz. Each with 22 MHz bandwidth. However, the successive channels overlap since the central frequencies of channels are spaced from each other by only 5 MHz. This causes channels 1, 6 and 11 to not overlap (see Table 7.1). Indeed, the central frequency of the channel 6 is: 2,412 + 5 x 5 = 2,437 MHz and the central frequency of the channel 11 is: 2,412 + 10 x 5 = 2,462 MHz.

Finally, in Table 7.2, we summarize the main characteristics of digital radio communications standards: UMTS (3G), LTE (4G) and IMT-2020 (5G). The IMT-2020 (*International Mobile Telecommunications*) standard uses the SCMA (*Sparse Code Multiple Access*) multiple access technique.

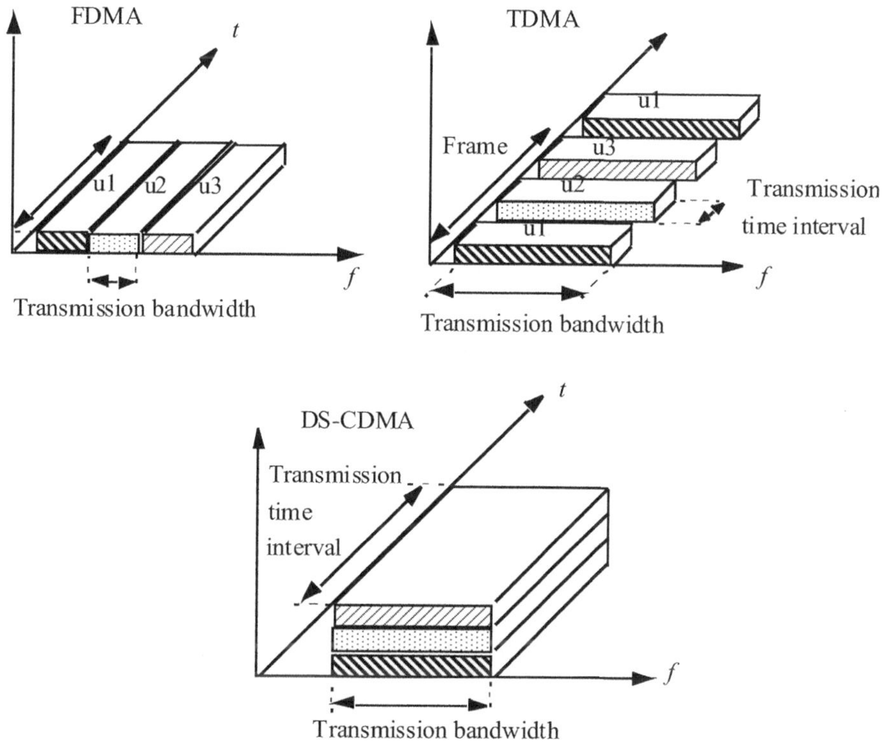

Figure 7.2. *Principle of multiple access techniques: FDMA, TDMA and DS-CDMA*

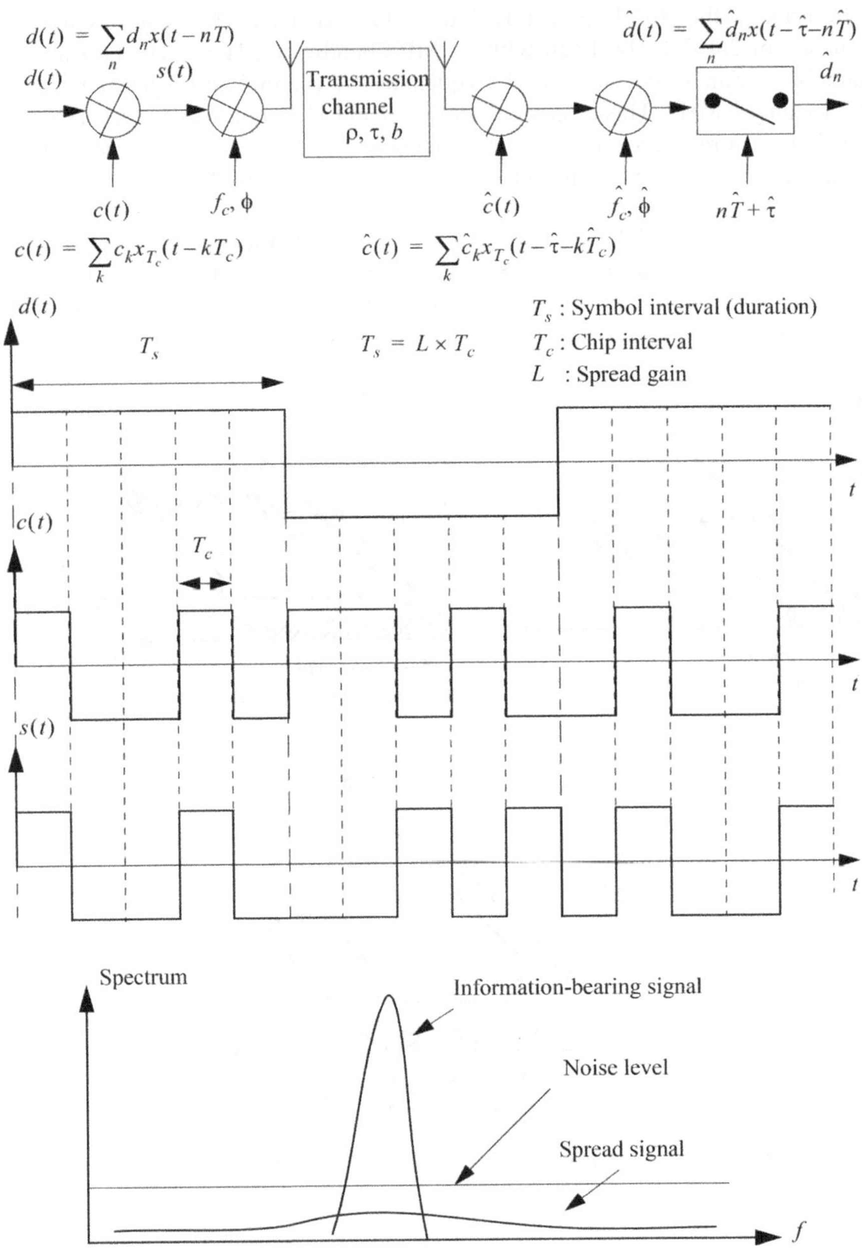

Figure 7.3. *Principle of direct sequence spread spectrum for a user*

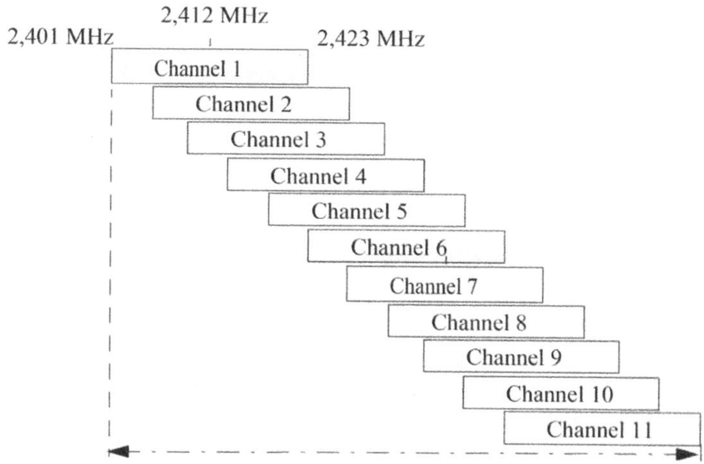

Figure 7.4. *Allocation in France of the OFDM spectrum in the 802.11.g standard*

Channel (MHz)	Low frequency	Central frequency	High frequency
Channel 1	2,401	2,412	2,423
Channel 6	2,426	2,437	2,448
Channel 11	2,451	2,462	2,473

Table 7.1. *Frequency characteristics of channels 1, 6 and 11*

	UMTS	LTE	IMT-2020
Bitrate (Mbit/s)	Uplink: 0.768 Downlink: 2	Uplink: 75 Downlink: 300	Downlink: 20,000
Frequency band (MHz)	Uplink: 1.92–1.9 Downlink: 2.11–2.17	1.4–20	3,460–3,800
Modulation	Uplink: BPSK Downlink: QPSK	Uplink et downlink: QPSK, 16-QAM 64-QAM	Filtered OFDM
Access type	CDMA	OFDMA	SCMA

Table 7.2. *Key features of the standards: UMTS (3G), LTE (4G) and IMT-2020 (5G)*

7.3. Structure of a radio link, a satellite link and a mobile radio channel

Here, we briefly describe three structures using the propagation of electromagnetic waves: the structure of a radio link and a satellite link, then the propagation through a radio-mobile channel in an urban environment in the case of a downlink.

7.3.1. *Structure of a terrestrial link (one jump)*

A radio-relay system is of the "*pseudo 4-frequency wires*". It uses two unidirectional channels using the same hardware support, but separated by different frequency bands, as shown in Figure 7.5. The link therefore comprises in each direction of transmission a transmitter E and a receiver R as well as a modulator M and a demodulator D.

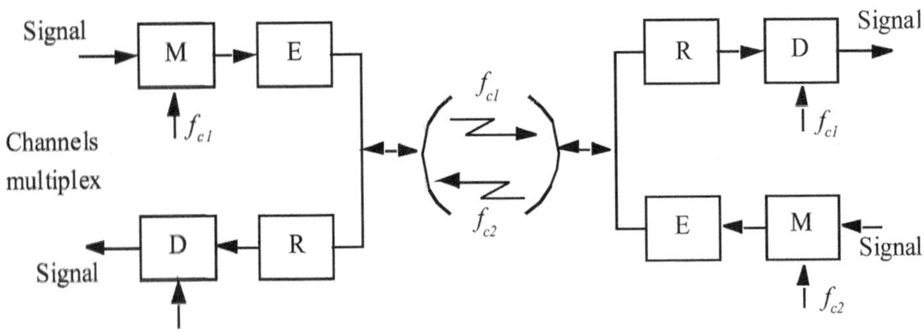

Figure 7.5. *Structure of a terrestrial link (one jump)*

The modulator transposes the baseband signal (signal carrying the useful information) around a carrier frequency and the transmitter produces a power wave carrying the modulated signal through the atmosphere. The receiver and the demodulator then perform the reverse operation providing the useful baseband signal back to the user.

7.3.2. *Structure of a satellite telecommunication link*

This is similar to a two-section terrestrial hertzian link and the satellite works as a relay station, as shown in Figure 7.6. The satellite repeater provides signal regeneration and amplification. The capacity of a repeater (1,000 telephone channels for a bandwidth of 36 Mhz on Intelsat or Eutelsat) allows the flow of traffic between several pairs of stations, providing that one of the multiple access techniques is used (FDMA, TDMA, DS-CDMA).

Most satellites use the 6/4 GHz band:

– f_{cm} = 5.925 to 6.425 GHz: earth-to-satellite path;

– f_{cd} = 3.7 to 4.2 GHz: satellite-to-earth path.

Thus, there is an available band of 500 MHz in each direction.

Other transmission bands are also used:

– 14/11 GHz with an available band of 500 MHz;

– 30/20 GHz with an available band of 3,500 MHz.

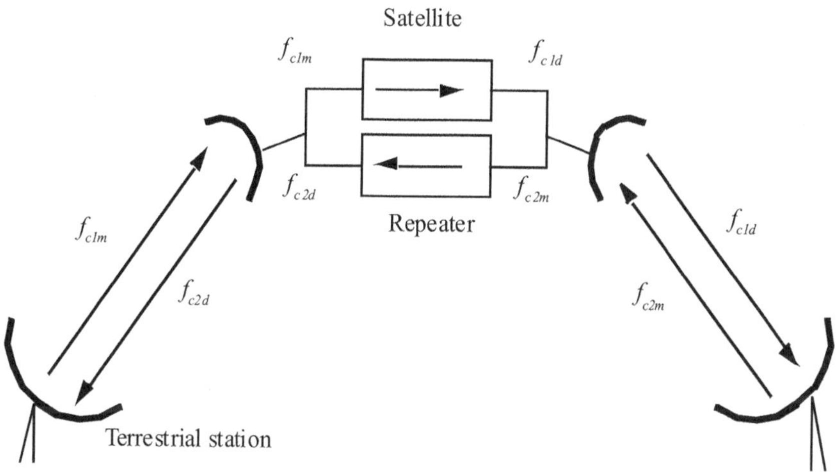

Figure 7.6. *Structure of a satellite link*

7.3.3. *Mobile radio channel*

The transmission of information by radio is done in both directions (duplex):

– from the base station (BS) to mobiles: *downlink*, as shown in Figure 7.7;

– from mobiles to the base station (BS): *uplink*.

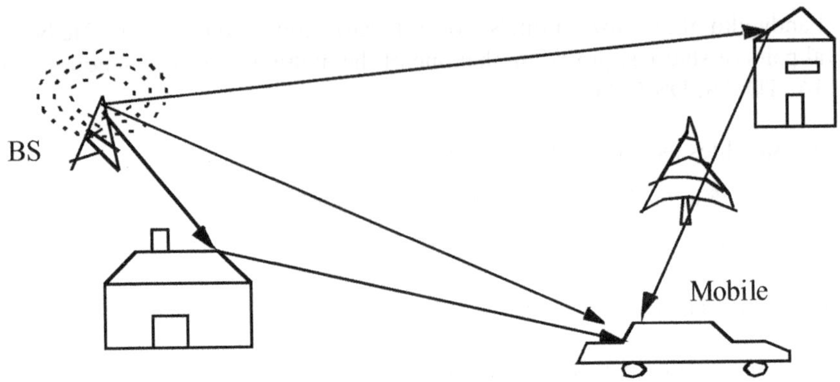

Figure 7.7. *Urban radio-mobile propagation: downlink*

The propagation conditions depend on the environment and the physical mechanisms of propagation that occur are:

– *reflexion*: occurs when an electromagnetic wave (EMW) encounters smooth surfaces (buildings, walls) of very large dimensions relative to the wavelength λ;

– *diffraction*: occurs when a thick and large obstacle compared to λ obstructs the EMW between the transmitter and the receiver. In this case, secondary EMWs are generated that propagate behind the obstacle (*shadowing*);

– *scattering*: occurs when the EMW encounters an obstacle whose thickness is of the order of the wavelength λ (rough surfaces, foliage, traffic lights). In this case, the energy is dispersed in all directions.

7.4. Effects of multiple paths and non-linearities of power amplifiers

7.4.1. *Effects of multiple paths: simple case of a direct path and only one delayed path*

In a radio-mobile or multi-path satellite transmission, the received signal can undergo frequency dispersions and time dispersions due to the Doppler effect. Two quantities are practically used to evaluate the behavior of a transmission in a multi-path environment. These quantities are the delay spread and the Doppler spread of the transmission channel.

Although Doppler spreading is a frequency-related phenomenon (offset of the carrier frequency), its effect on the received signal is manifested by a selective behavior in the time domain.

Similarly, delay spread of the received signal, which is a temporal phenomenon, is manifested by a selective behavior in the frequency domain (Assaf *et al.* 2010; Pätzold 2002).

In the following, we analyze a simple case: reception of a direct path added to a single delayed path (see Figure 7.8). This simple case will allow us to demonstrate the fluctuation of the signal received during a communication between two users in motion.

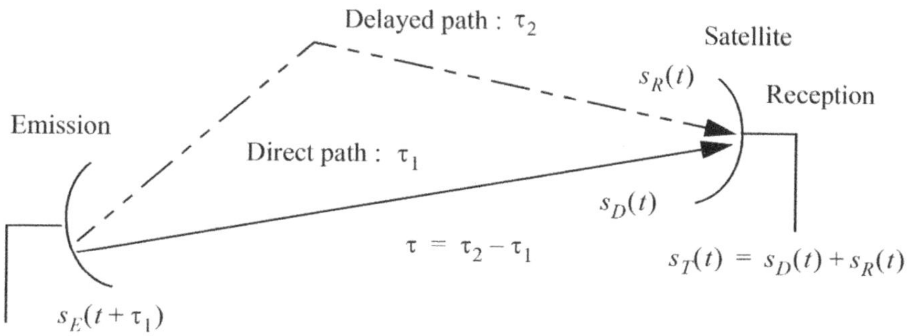

Figure 7.8. *Direct and delayed propagation in the case of a terrestrial-satellite transmission*

Suppose that the signal emitted $s_E(t+\tau_1)$ at time $(t+\tau_1)$ arrives by a direct path after a delay τ_1 at the receiver. This signal is called $s_D(t)$. Also, this same emitted signal arrives at the receiver after having undergone a reflection and delay τ_2. This second signal is called $s_R(t)$. Examine now the total signal received $s_T(t)$ considering the relative delay $\tau = \tau_2 - \tau_1$.

The direct signal received can be written:

$$s_D(t) = s_{D,e}(t)\exp(j2\pi f_c t) \quad [7.1]$$

where $s_{D,e}(t)$ is the complex envelope of the signal $s_D(t)$. It represents the informative baseband signal (we will see the justification of this concept later in this chapter).

The delayed signal received $s_R(t)$ is then written:

$$s_R(t) = \rho s_{D,e}(t-\tau)\exp[j2\pi f_c(t-\tau)] \quad [7.2]$$

with:

$$\rho = |\rho|\exp(j\varphi) \quad [7.3]$$

the complex reflection coefficient (attenuation).

The total signal received is:

$$\begin{aligned} s_T(t) &= \left[s_{D,e}(t) + |\rho|\exp[j(\varphi - 2\pi f_c\tau)]s_{D,e}(t-\tau)\right]\exp(j2\pi f_c t) \\ &= s_{T,e}(t)\exp(j2\pi f_c t) \end{aligned} \quad [7.4]$$

where $s_{T,e}(t)$ is the complex envelope (equivalent baseband signal) of $s_T(t)$.

In the frequency domain, the total received signal $S_{T,e}(f)$ is obtained by the Fourier transform of $s_{T,e}(t)$ (see relation [7.4]):

$$\begin{aligned} S_{T,e}(f) &= S_{D,e}(f)[1 + |\rho|\exp[j(\varphi - 2\pi f_c\tau)]\exp(-j2\pi f\tau)] \\ &= S_{D,e}(f) \times M(f) \end{aligned} \quad [7.5]$$

The term $M(f)$ oscillates between a maximum value and a minimum value depending on the value of the phase as clearly indicated by relations [7.6] and [7.7] respectively:

$$M_{max}(f) = 1 + |\rho| \quad \text{for} \quad \varphi - 2\pi\tau[f_c + f] = \pm 2k\pi \quad [7.6]$$

$$M_{min}(f) = 1 - |\rho| \quad \text{for} \quad \varphi - 2\pi\tau[f_c + f] = \pm(2k+1)\pi \quad [7.7]$$

The term $1 - |\rho|$ is called notch depth.

The first minimum is observed for a frequency f_0 such that: $\varphi - 2\pi\tau[f_c + f] = \pi$, then for all frequencies of the form:

$$f_k = f_0 + \frac{k}{\tau} \quad [7.8]$$

The first maximum is observed for a frequency f_0' such that $\varphi - 2\pi\tau[f_c + f_0'] = 0$, then for all frequencies of the form: $f_k' = f_0' + \frac{k}{\tau}$

Frequencies f_0 and f_0' are a function of φ, τ, f_c and $f_0 = f_0' + \frac{1}{2\tau}$

In Figure 7.9, we have plotted the term $M(f)$ for a relative delay $\tau = 50$ ns (usual case in earth-satellite transmission) and a reflection coefficient $|\rho| = 0.9$.

$$1/\tau = 20 \text{ MHz}\,;\ |\rho| = 0.9\ ;\ 1 - |\rho| = -20 \text{ dB}\,;\ 1 + |\rho| = 5.57 \text{ dB}$$

Thus, it is observed that the action of the delayed signal added to that of the direct path causes a modulation of the gain in the frequency domain, amplifying it on certain frequency domains and attenuating it on others (see Figure 7.9).

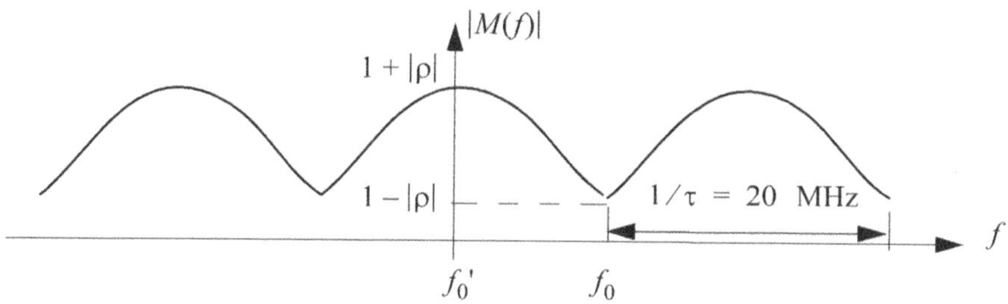

Figure 7.9. *Oscillation of |M(f)| for parameters typical of a terrestrial-satellite transmission*

7.4.2. *Effects of non-linearities of power amplifiers*

In Figure 7.10, we show the detrimental effects of the non-linearities of the power amplifiers. Indeed, from a certain value of the input signal amplitude of the power amplifier (threshold), we notice:

– the appearance of secondary lobes in the output signal spectrum of the power amplifier which will be transformed into a wave emitted by the transmitting antenna;

– the collapse of the amplitude (backoff) and phase rotation on the output signal of the power amplifier. All this has a direct impact on the constellation diagram (we considered QPSK modulation to illustrate this phenomenon) and generates intersymbol interference. The higher the non-linearity, the stronger the observed phenomenon.

Finally, we note that it is more difficult to combat the effects of the non-linearities of power amplifiers compared to the effects of multiple paths.

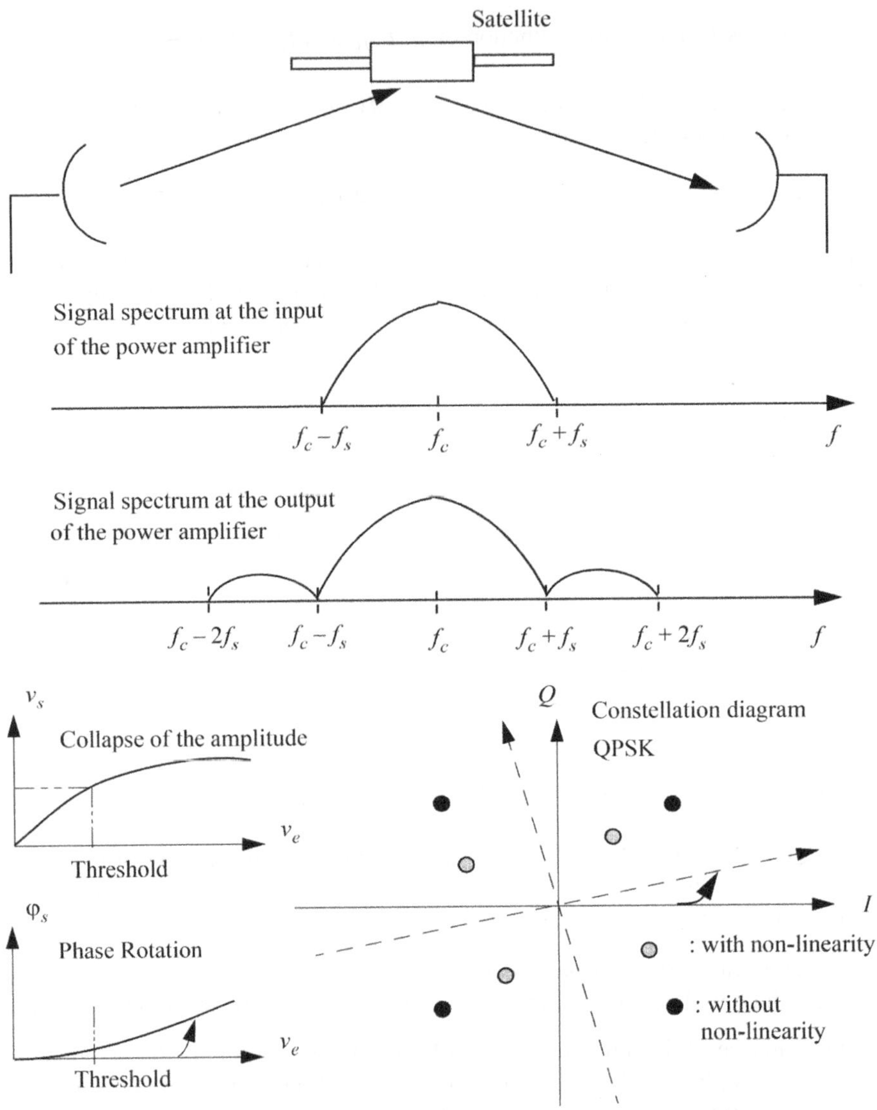

Figure 7.10. *Harmful effects of non-linearities of power amplifiers on transmission*

7.5. Linear digital carrier modulations

7.5.1. *Principle*

In linear digital modulation of a carrier, the modulated signal $s(t)$ is obtained by the discrete variation, according to the M possible values of the modulating baseband signal $c(t)$ (information-bearing signal, generally complex) of frequency bandwidth B, of one of the parameters, amplitude V or phase φ_c (or both) of the carrier signal $p_c(t)$ which is written:

$$p_c(t) = V\exp[j(2\pi f_c t + \varphi_c)] \qquad [7.9]$$

with $f_c > B$ and f_c is the carrier frequency.

The signal $c(t)$ is given by the following relation:

$$c(t) = \sum_n c_n\, x(t - nT) = \sum_n \rho_n \exp(j\psi_n) x(t - nT) \qquad [7.10]$$

with:

$$c_n = a_n + ja'_n = \rho_n \exp(j\psi_n) \qquad [7.11]$$

and:

$$x(t) = \begin{cases} 1 & \text{if } t \in [-T/2, T/2[\\ 0 & \text{elsewhere} \end{cases} \qquad [7.12]$$

the shape of the electric pulse emitted.

The modulated signal is then written as being the real part of the product of the modulating signal by the carrier signal:

$$s(t) = \Re\{c(t)V\exp[j(2\pi f_c t + \varphi_c)]\} \qquad [7.13]$$

If the discrete variation according to the M possible values of the modulating signal concerns both the amplitude and phase parameters, this is called *amplitude and phase modulation* (polar space), or *amplitude modulation of two quadrature carriers* (Cartesian space).

7.5.2. General characteristics of the modulated signal s(t)

We will determine the second-order statistical properties of the signal $s(t)$.

7.5.2.1. *Mean value*

Since the carrier is a deterministic function of time, then the first statistical moment of $s(t)$ is:

$$E[s(t)] = \Re\{E[c(t)] \times V exp[j(2\pi f_c t + \varphi_c)]\} \quad [7.14]$$

We see that if $E[c(t)]$ is non-zero, then $E[s(t)]$ is a periodic function of period $1/f_c$. So, $s(t)$ is cyclostationary of the first order.

The statistical mean $E[c(t)]$ of the signal $c(t)$ is:

$$E[c(t)] = (m_a + jm_{a'}) \sum_n x(t - nT) \quad [7.15]$$

It is periodic of period T. m_a, $m_{a'}$ respectively denote the averages of the symbols a and a'.

7.5.2.2. *Autocorrelation function*

The autocorrelation of the modulated signal $s(t)$ is defined by the following relation:

$$\begin{aligned} R_{ss}(t, t-\tau) &= E[s(t)s^*(t-\tau)] \\ &= \frac{V^2}{2}\Re\{E[c(t)c^*(t-\tau)] \times \exp[j(2\pi f_c t + \varphi_c)] \\ &\times \exp[-j[2\pi f_c(t-\tau) + \varphi_c]]\} \end{aligned} \quad [7.16]$$

where $\Re, *$ signifies a real and complex conjugate respectively.

The previous relationship becomes:

$$R_{ss}(t, t-\tau) = E[s(t)s^*(t-\tau)] = \frac{V^2}{2}\Re\{R_{cc}(t, t-\tau)\exp(j2\pi f_c \tau)\} \quad [7.17]$$

where $R_{cc}(t, t-\tau)$ denotes the autocorrelation function of the baseband signal $c(t)$.

Relation [7.17] shows that the autocorrelation function $R_{ss}(t, t-\tau)$ is real and even, but remains by function $R_{cc}(t, t-\tau)$ dependent on t and $t-\tau$. The autocorrelation function $R_{cc}(t, t-\tau)$, after development, is given by:

$$\begin{aligned} R_{cc}(t, t-\tau) &= R_{cc}(t+\tau, t) \\ &= \sum_k R_{AA}(k) \sum_n x(t-nT) \times x(t-nT-kT+\tau) \end{aligned} \qquad [7.18]$$

where $R_{AA}(k)$ is the autocorrelation function of complex symbols $A_k = c_k = a_k + ja_k^{'}$ assumed to be stationary.

The second term of relation [7.18] is periodic by the variable t, of period T. So, the signal $c(t)$ is, in the broader sense, a cyclostationary random function (because $E[c(t)]$ and $R_{cc}(t+\tau, t)$ are periodic of period T). We can then replace $R_{cc}(t+\tau, t)$ by its average value over a period T. This makes it possible to calculate the power spectral density $\Gamma_c(f)$.

We note, $R_{cc}(\tau)$ the average over a period T of $R_{cc}(t+\tau, t)$. We show that:

$$R_{cc}(\tau) = \frac{1}{T}\int_0^T R_{cc}(t+\tau, t)dt = \frac{1}{T}[R_{AA}(k) \otimes R_{xx}(\tau)] \qquad [7.19]$$

However, if the symbols a_n are independent and likewise for the symbols $a_n^{'}$ and if, moreover, the symbols a_n and $a_n^{'}$ are independent of each other, then the autocorrelation function $R_{AA}(k)$ is:

$$R_{AA}(k) = \begin{cases} [\sigma_a^2 + m_a^2] + [\sigma_{a'}^2 + m_{a'}^2] & \text{if } k = 0 \\ [m_a^2 + m_{a'}^2] & \text{if } k \neq 0 \end{cases} \qquad [7.20]$$

7.5.2.3. *Power spectral density*

The power spectral density of the complex symbols $A_k = c_k = a_k + ja_k^{'}$ is the discrete Fourier transform (DFT) of $R_{AA}(k)$. We then obtain:

$$\Gamma_A(f) = DFT\{R_{AA}(k)\} = [\sigma_a^2 + \sigma_{a'}^2] + [m_a^2 + m_{a'}^2] \times \frac{1}{T}\sum_{k=-\infty}^{\infty} \delta\left(f - \frac{k}{T}\right) \qquad [7.21]$$

where $\sigma_a^2 + \sigma_{a'}^2$ is the variance of the random variable: $A_k = c_k = a_k + ja_k^{'}$ and $[m_a^2 + m_{a'}^2]^{1/2}$ is the modulus of its mean value.

The power spectral density $\Gamma_c(f)$ of the signal $c(t)$, according to relation [7.19], is:

$$\Gamma_c(f) = F\{R_{cc}(\tau)\} = \frac{1}{T}[\Gamma_A(f) \times \Gamma_x(f)] = \frac{1}{T}[\Gamma_A(f) \times |X(f)|^2] \qquad [7.22]$$

that is, more specifically:

$$\Gamma_c(f) = \frac{[\sigma_a^2 + \sigma_{a'}^2]}{T}|X(f)|^2 + \frac{[m_a^2 + m_{a'}^2]}{T^2}\sum_{k=-\infty}^{\infty}\left|X\left(\frac{k}{T}\right)\right|^2 \delta\left(f - \frac{k}{T}\right) \qquad [7.23]$$

It is composed of a continuous spectrum (which always exists because σ_a^2 and $\sigma_{a'}^2$ are positive) and of discrete spectral components. This latter is zero if the means m_a and $m_{a'}$ are zero.

As $R_{cc}(t+\tau, t)$ is periodic of period T, $R_{ss}(t+\tau, t)$, according to [7.17], is also periodic of period T. Therefore, $s(t)$ is a 2nd order cyclostationary signal.

We can then replace $R_{ss}(t+\tau, t)$ given by [7.17] by its average over a period T, noted $R_{ss}(\tau)$:

$$R_{ss}(\tau) = \frac{1}{T}\int_0^T R_{ss}(t+\tau, t)dt = \Re\left\{\frac{V^2}{2}\exp(j2\pi f_c\tau) \times R_{cc}(\tau)\right\} \qquad [7.24]$$

The power spectral density $\Gamma_s(f)$ is calculated as follows. Let:

$$z(\tau) = \frac{V^2}{2}\exp(j2\pi f_c\tau) \times R_{cc}(\tau) = x(\tau) + y(\tau) \Rightarrow \Re[z(\tau)] = x(\tau)$$
$$= \frac{1}{2}[z(\tau) + z^*(\tau)] \qquad [7.25]$$

hence:

$$\Gamma_s(f) = F\{R_{ss}(\tau)\} = \frac{V^2}{4} \times F\{R_{cc}(\tau)\exp(j2\pi f_c\tau) + R^*{}_{cc}(\tau)\exp(-j2\pi f_c\tau)\}$$

$$\Gamma_s(f) = \frac{V^2}{4}[\Gamma_c(f) \otimes \delta(f - f_c) + \Gamma_c{}^*(-f) \otimes \delta(-f - f_c)]$$

$$\Gamma_s(f) = \frac{V^2}{4}[\Gamma_c(f - f_c) + \Gamma_c{}^*(-f - f_c)] \qquad [7.26]$$

Relation [7.26] and Figure 7.11 clearly show that the frequency band occupied by the modulated signal $s(t)$ is the frequency band B of the modulating signal $c(t)$ transposed at frequencies f_c and $-f_c$.

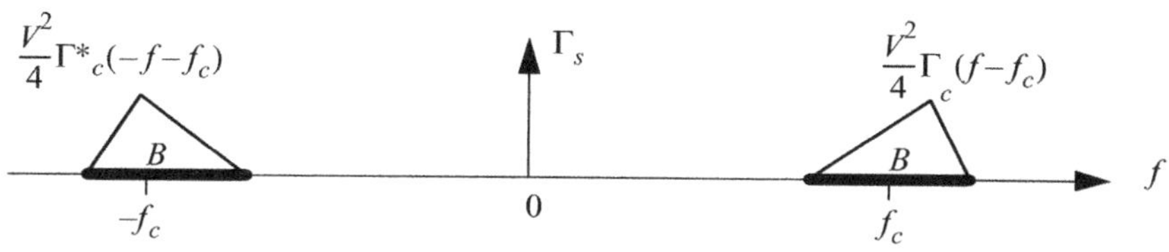

Figure 7.11. *Power spectral density of the modulated signal*

7.6. Quadrature digital linear modulations: general structure of the modulator, spatial diagram, constellation diagram and choice of a constellation

From relations [7.13], [7.10] and [7.11], which we recall respectively hereinafter:

$$s(t) = \Re\{c(t)V\exp[j(2\pi f_c t + \varphi_c)]\}$$

$$c(t) = \sum_n c_n\, x(t - nT) = \sum_n \rho_n \exp(j\psi_n)x(t - nT)$$

$$c_n = a_n + ja'_n = \rho_n \exp(j\psi_n)$$

we can obtain a more general form of the modulated signal which reveals the combined modulation of amplitude and phase:

$$\begin{cases} s(t) = \Re\left\{V\sum_n \rho_n \exp(j\psi_n)x(t - nT)\exp[j(2\pi f_c t + \varphi_c)]\right\} \\ s(t) = \Re\left\{V\sum_n \rho_n x(t - nT)\exp[j(2\pi f_c t + \varphi_c + \psi_n)]\right\} \\ s(t) = V\sum_n \rho_n x(t - nT)\cos(2\pi f_c t + \varphi_c + \psi_n) \end{cases} \quad [7.27]$$

with:

$$\rho_n = \left[a_n^2 + a_n'^2\right]^{1/2} \text{ and } \psi_n = \tan^{-1}\left[\frac{a_n'}{a_n}\right] \quad [7.28]$$

Or, by developing:

$$s(t) = V\sum_n \rho_n \cos(\psi_n) x(t - nT)\cos(2\pi f_c t + \varphi_c)$$
$$-V\sum_n \rho_n \sin(\psi_n) x(t - nT)\sin(2\pi f_c t + \varphi_c) \quad [7.29]$$

where $[\rho_n, \psi_n]$ belong to the set of M possible amplitude-phase pairs, expressing the amplitude and phase states in the time interval T.

Moreover, as:

$$a_n = \rho_n \cos(\psi_n) \text{ and } a_n' = \rho_n \sin(\psi_n) \quad [7.30]$$

Relation [7.29] can be written:

$$s(t) = V\sum_n a_n\, x(t - nT)\cos(2\pi f_c t + \varphi_c)$$
$$-V\sum_n a_n'\, x(t - nT)\sin(2\pi f_c t + \varphi_c) \quad [7.31]$$

with a_n being $M1$-ary and a_n' being $M2$-ary.

Note then that [7.29] is formally identical to relation [7.31]. However, while in quadrature amplitude modulation with $M = M1 \times M2$ states uses cartesian representation separability, the joint amplitude and phase shift keying modulation with $M = M1 \times M2$ states uses polar representation separability.

7.6.1. *General structure of the modulator*

Relation [7.31] can also be written:

$$s(t) = V[I(t)\cos(2\pi f_c t + \varphi_c) - Q(t)\sin(2\pi f_c t + \varphi_c)] \quad [7.32]$$

with:

$$I(t) = \sum_n a_n x(t - nT) \quad \text{and} \quad Q(t) = \sum_n a'_n x(t - nT) \qquad [7.33]$$

The signals $I(t)$ and $Q(t)$ are linear combinations of the digital baseband signals. They modulate in amplitude the in-phase carrier $\cos(2\pi f_c t + \varphi_c)$ and the in-quadrature carrier $-\sin(2\pi f_c t + \varphi_c)$. Hence the general structure of the modulator given in Figure 7.12 below.

We have added the band-pass filter (BPF) $g_e(t)$ centered around the frequency f_c which allows, on the one hand, to limit the spectrum of the baseband signals $I(t)$ and $Q(t)$ and, on the other hand, to ensure the optimization of all the elements of the communication chain (optimal filtering distribution between transmission and reception).

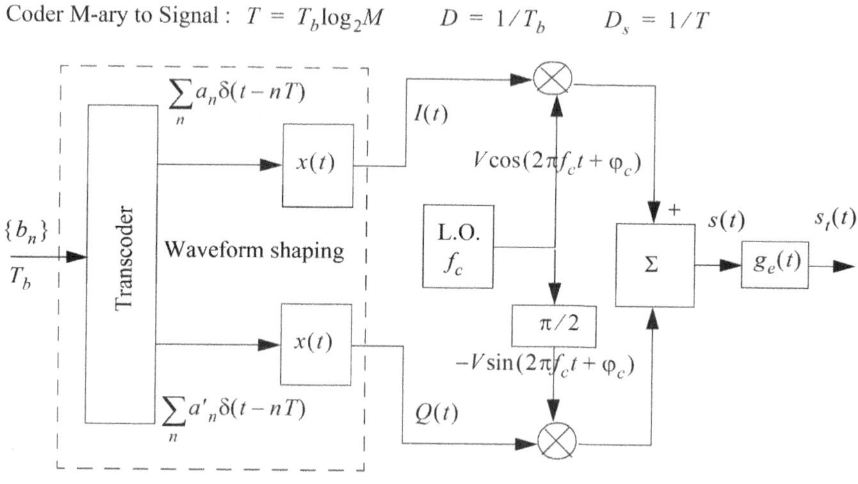

Figure 7.12. *General structure of the linear digital modulator. For a color version of this figure, see www.iste.co.uk/assad/digital1.zip*

7.6.2. *Spatial diagram (or vectorial) and constellation diagram*

Relation [7.32] can also be written:

$$s(t) = V\big[\rho(t)\cos\big(2\pi f_c t + \varphi_c + \psi(t)\big)\big] \qquad [7.34]$$

with:

$$\rho(t) = [I^2(t) + Q^2(t)]^{1/2} \text{ and } \psi(t) = \tan^{-1}\left[\frac{Q(t)}{I(t)}\right] \quad [7.35]$$

To the signal $s(t)$ given by [7.32] and [7.34], we can associate a vector $\overrightarrow{OS}$ in a two-dimensional space, with as basic vectors:

$$e_1(t) = VI(t)\cos(2\pi f_c t + \varphi_c) \text{ and } e_2(t) = VQ(t)\sin(2\pi f_c t + \varphi_c) \quad [7.36]$$

therefore, the vector $\overrightarrow{OS}$:

$$\overrightarrow{OS} = \begin{cases} |\overrightarrow{OS}| = V\rho(t) \\ Arg(\overrightarrow{OS}) = \psi(t) \end{cases} \quad [7.37]$$

globally describes the amplitude and the phase of $s(t)$ in the $e_1 0 e_2$ plan at any time (see Figure 7.13) and thus forms the spatial diagram or vectorial diagram (see Chapter 6).

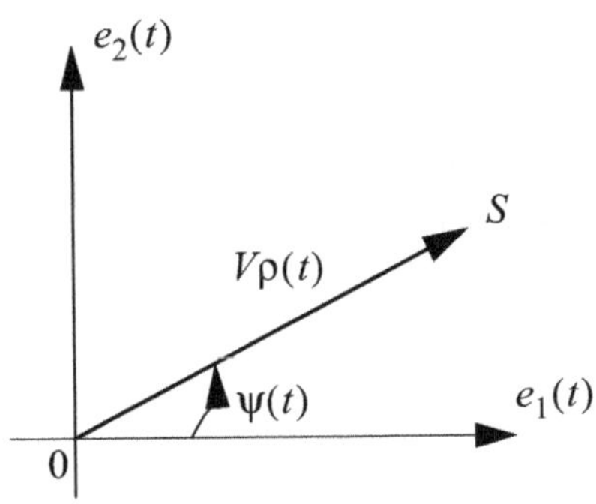

Figure 7.13. *Consrtuction of the spatial diagram*

In digital transmission, the useful information is contained in the sampled values of the vector $\overrightarrow{OS}$ at times of the form $t_k = kT$.

The *locus* of the vector $\overrightarrow{OS}$ at the sampling instants is a *constellation of points* $\begin{bmatrix} V\rho(kT) \\ \psi(kT) \end{bmatrix}$ called a *constellation diagram*.

NOTE.– In the constellation diagram, it has been admitted that $X(f)$ has a limited frequency support $X(f) \in]-f_c, f_c[$.

Indeed, if the product $f_c T$ is an integer, then the scalar product is:

$$\langle \cos(2\pi f_c t + \varphi_c), \sin(2\pi f_c t + \varphi_c) \rangle = 0$$

When $X(f)$ is not with a bounded support, in the case of the NRZ signal for example, and because:

$$|X(f)|^2 = \left[\frac{\sin(\pi f T)}{\pi f T}\right]^2$$

then to preserve the vector representation, it is necessary to impose that: $f_c T \gg 1$.

In practice, the transmission filter limits the frequency band of $X(f)$ to: $X(f) \in]-B, B[$, hence, it is sufficient to use the narrow band hypothesis: $f_c \gg B$.

7.6.3. *Choosing a constellation diagram*

The choice of a constellation diagram depends on:

– the probability of error P_e;

– the complexity of the modem;

– the symbol rate (or modulation speed).

In the hypothesis of independent symbols, the probability of error P_e depends on:

– the minimal Euclidean distance between states:

$$d_{min} = \underbrace{min}_{i,j} \left|\overrightarrow{OS_j} - \overrightarrow{OS_i}\right|_{i \neq j} \quad [7.38]$$

– the maximum possible peak amplitude of the modulated signal:

$$d_{Max} = \underbrace{Max}_{i} \left|\overrightarrow{OS_i}\right| \quad [7.39]$$

– the noise power.

7.7. Digital radio transmission and equivalent baseband digital transmission: complex envelope

In this section, we first give the general block diagram of a digital radio transmission including the different signals of the different elements constituting it. Then, we present the simplified block diagram grouping all the filters of the transmission chain into a single global filter $H(f)$.

This approach will then enable us to switch from a digital radio transmission to an equivalent digital baseband transmission, thanks to the notion of the equivalent baseband filter $H_e(f)$ to the global filter $H(f)$ and to the notion of the complex envelope of the band-pass signals. This will later allow the use of the same tools and parameters (eye diagram, constellation diagram, bit error probability, etc.) to evaluate the performance of the digital radio transmission chain.

The general scheme of a digital radio transmission system is given in Figure 7.14 below.

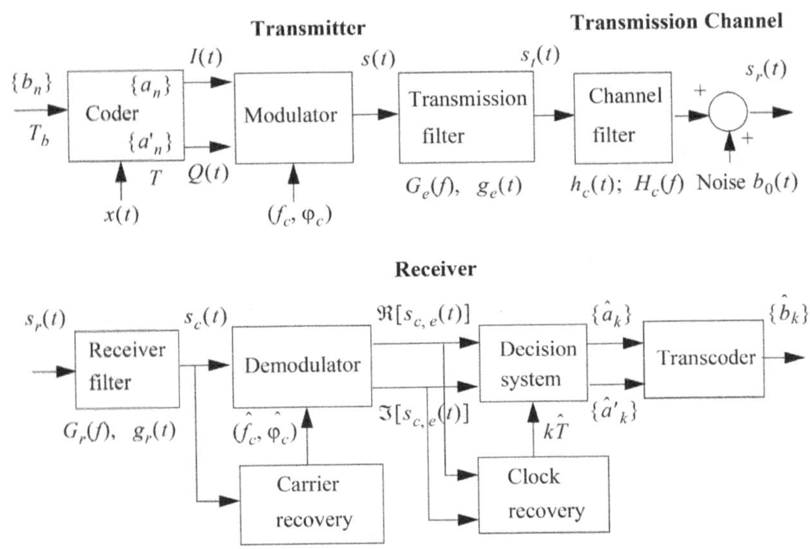

Figure 7.14. *General scheme of a digital radio transmission system*

Now let's replace all the filters in the digital radio transmission chain with the filter $H(f)$:

$$H(f) = G_e(f) \times H_c(f) \times G_r(f) \quad [7.40]$$

whose impulse response is:

$$h(t) = g_e(t) \otimes h_c(t) \otimes g_r(t) \tag{7.41}$$

The simplified schematic block diagram of a digital radio transmission system is shown in Figure 7.15.

In a quadrature amplitude modulation (QAM), the signal transmitted $s(t)$ is of the form:

$$s(t) = \Re\left\{V \sum_n [a_n + ja'_n]\, x(t - nT) \exp[j(2\pi f_c t + \varphi_c)]\right\} \tag{7.42}$$

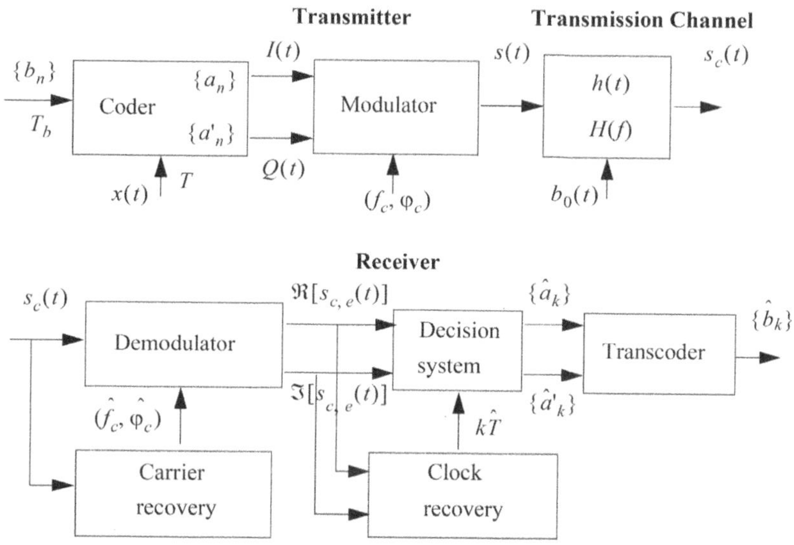

Figure 7.15. *Simplified block diagram of a digital radio transmission system*

7.7.1. *Equivalent baseband digital transmission: complex envelope*

We now pose an important question: how can one consider a digital radio transmission (bandpass transmission) as a baseband equivalent digital transmission?

For this purpose, it is sufficient to replace the different signals with a carrier by their complex envelope, and the global real bandpass filter $H(f)$ around the frequencies f_c and $-f_c$, by its equivalent baseband filter $H_e(f)$, which is a complex one, as shown in Figure 7.16.

Later, we will see all the justifications proving the accuracy of the equivalent baseband transmission model and its interest.

We denote $y_e(t)$ the response of the filter $h_e(t)$ to the electric signal $x(t)$. Its general form is:

$$y_e(t) = p(t) + jq(t) \quad [7.43]$$

It is complex because the filter $h_e(t)$ is complex.

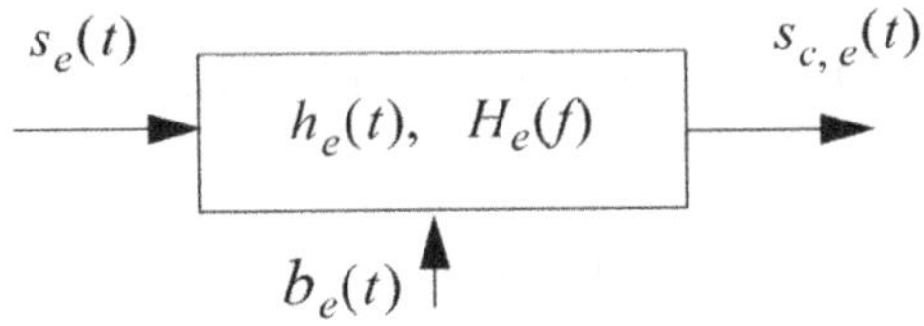

Figure 7.16. *Complex envelope and equivalent baseband filter*

In addition, the result of noise filtering $b_0(t)$ by the bandpass filter of the receiver (impulse response $g_r(t)$), called noise $b(t)$, has a complex envelope $b_e(t)$ of the following form:

$$b_e(t) = b_{e,p}(t) + jb_{e,q}(t) \quad [7.44]$$

Under these conditions, the response of the filter $h_e(t)$ to the complex envelope $s_e(t)$ of the transmitted signal $s(t)$, with:

$$s_e(t) = V\sum_n [a_n + ja'_n]\, x(t - nT) = V[I(t) + jQ(t)] \quad [7.45]$$

and so:

$$s_{c,e}(t) = V\sum_n [a_n + ja'_n]\, [p(t - nT) + jq(t - nT)] + b_e(t) \quad [7.46]$$

At sampling times $t_k = kT$ (neglecting time t_0, see Chapter 6), expression [7.46] is written:

$$s_{c,e}(kT) = V\sum_n [a_n + ja'_n]\, \{p[(k - n)T] + jq[(k - n)T]\} + b_e(kT) \quad [7.47]$$

The real part of $s_{c,e}(kT)$ allows the decision system (see Figure 7.15) to estimate the transmitted symbol a_k, i.e. $\hat{a}_k$. The imaginary part allows the decision system to estimate the transmitted symbol $a_k^{'}$, i.e. $\hat{a}_k^{'}$.

We consider now only the real part of $s_{c,e}(kT)$. By isolating in the summation the term that contains the useful transmitted information a_k, it is written thus:

$$\Re[s_{c,e}(kT)] = V\left\{a_k p(0) + \sum_{n\neq k} a_n p[(k-n)T] - \sum_{n} a_n' q[(k-n)T]\right\} \quad [7.48]$$
$$+b_{e,p}(kT)$$

It is similar to expression [6.15] from Chapter 6, and also contains three terms:

– the useful signal $Va_k p(0)$;

– the sum of the other answers:

$$I_{m_I}(kT) = V\left[\underbrace{\sum_{n\neq k} a_n p[(k-n)T]}_{\text{intra-channel ISI}} - \underbrace{\sum_{n} a_n' q[(k-n)T]}_{\text{inter-channel ISI}}\right] \quad [7.49]$$

which is an interymbol interference (ISI), itself composed of two parts: the first part represents the *intra-channel ISI* (as in the case of a baseband transmission, see Chapter 6, relationship [6.16]) and the second part represents the *inter-channel ISI* which does not exist in the case of a baseband transmission, and that takes into account all symbols $a_n^{'}$;

– the noise $b_{e,p}(kT)$.

We consider now only the imaginary part of $s_{c,e}(kT)$. By isolating in the summation the term $a_k^{'}$ which contains the useful transmitted information, it is written:

$$\Im[s_{c,e}(kT)] = V\left\{a_k' p(0) + \sum_{n\neq k} a_n' p[(k-n)T] + \sum_{n} a_n q[(k-n)T]\right\} \quad [7.50]$$
$$+b_{e,q}(kT)$$

It is also composed of three terms:

– the useful signal $Va_k' p(0)$;

– the sum of the other answers:

$$I'_{m'_l}(kT) = V\left[\underbrace{\sum_{n\neq k} a'_n p[(k-n)T]}_{\text{intra-channel ISI}} + \underbrace{\sum_{n} a_n q[(k-n)T]}_{\text{inter-channel ISI}}\right] \quad [7.51]$$

which is an intersymbol interference (ISI), itself also composed of two parts: the first part represents the *intra-channel ISI* and the second part represents the *inter-channel ISI*;

– the noise $b_{e,q}(kT)$.

Now that we know the answer of $s_{c,e}(t)$, we can deduce that of $s_c(t)$ thanks to the definition of the analytic signal.

Indeed, we can write:

$$s_c(t) = \Re\{s_{c,e}(t) \times \exp[j(2\pi f_c t + \varphi_c)]\} \quad [7.52]$$

after development, we get:

$$s_c(t) = V\sum_n [a_n p(t-nT) - a'_n q(t-nT)]\cos(2\pi f_c t + \varphi_c)$$
$$-V\sum_n [a'_n p(t-nT) + a_n q(t-nT)]\sin(2\pi f_c t + \varphi_c) + b(t) \quad [7.53]$$

Figure 7.17 shows the distribution of the signals on the in-phase carrier and on the in-quadrature carrier.

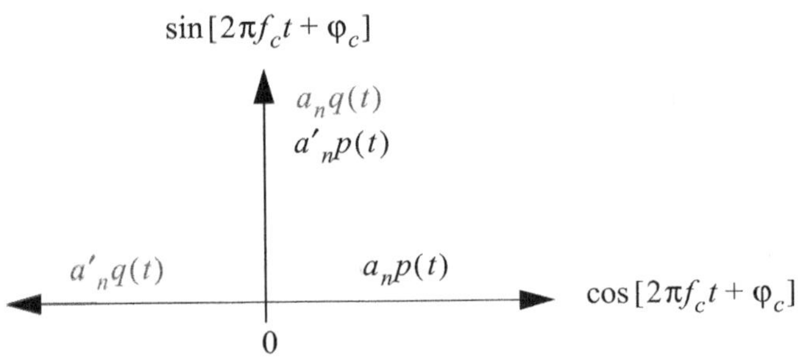

Figure 7.17. *Distribution of the signals on the carrier in-phase and on the carrier in-quadrature. For a color version of this figure, see www.iste.co.uk/assad/digital1.zip*

7.8. Equivalent baseband transmission, interest and justification: analytical signal and complex envelope

In the previous section, we have already shown a primary interest of the equivalent baseband digital transmission model that allows a relatively simple analytical study and easy evaluation via the tools and parameters developed in Chapter 6 on baseband transmission. In the following, we will show a second interest related to the simulation time, then the theoritical justification of this model will be given.

7.8.1. *Interest: important simplification in numerical simulation*

We will demonstrate this aspect via the following example. Let:

– a baseband signal (speech signal for example) of bandwidth $B = 4\text{KHz}$, hence a symbol sampling frequency $f_s = 2B$ and a symbol timeslot of $T_s = 1/f_s = 125 \times 10^{-6}\ s$;

– a carrier frequency $f_c = 2\text{GHz}$, of period: $T_c = 1/f_c = 5 \times 10^{-10}\ s$.

A symbol period (symbol timeslot) contains $T_s/T_c = \frac{1}{4} \times 10^6$, a quarter million carrier periods. To simulate a carrier period (RF), according to Shannon's theory, at least a sampling frequency $f_e \geq 2f_c$ is required. So, to simulate a symbol period, we have to use *1/2 million samples*. However, *only eight samples* were needed for simulating the baseband transmission of the speech signal. This demonstrates *an interest of the digital baseband transmission model without any loss of information (under certain constraints).*

7.8.2. *Analytical signal and complex envelope of a modulated signal*

Let a modulated signal $s(t)$, the analytical signal associated with $s(t)$, called $s_a(t)$ be rigorously defined by:

$$s_a(t) = s(t) + j\vec{s}(t) \qquad [7.54]$$

and conversely, the signal $s(t)$ is given by:

$$s(t) = \Re[s_a(t)] \qquad [7.55]$$

with, $\vec{s}(t)$ defined as the Hilbert transform of the signal $s(t)$:

$$\vec{s}(t) = \text{Hilbert}[s(t)] = s(t)\otimes\frac{1}{\pi t} = v.p\frac{1}{\pi}\int_{-\infty}^{\infty}\frac{s(\tau)}{t-\tau}d\tau \qquad [7.56]$$

where $v.p$ is the principal value of Cauchy.

This transformation is equivalent to a $\pi/2$ phase shift of the signal $s(t)$. Indeed, if:

$$\begin{cases} z = \cos(2\pi ft) & \rightarrow \text{Hilbert}[z] = \sin(2\pi ft) & \rightarrow z_a = \exp(j2\pi ft) \\ z = \sin(2\pi ft) & \rightarrow \text{Hilbert}[z] = -\cos(2\pi ft) & \rightarrow z_a = -j\exp(j2\pi ft) \end{cases} \qquad [7.57]$$

In the frequency domain, signals $\vec{S}(f)$ and $S_a(f)$ are expressed as a function of the signal $S(f)$ by:

$$\vec{S}(f) = -j\text{sign}(f) \times S(f) \qquad [7.58]$$

and:

$$S_a(f) = \begin{cases} 2S(f) & \text{for } f > 0 \\ S(f) & \text{for } f = 0 \\ 0 & \text{for } f < 0 \end{cases} \qquad [7.59]$$

Let's apply these results on the signal $s(t)$ modulated in M-QAM (or in M-PSK). This can be written as follows:

$$s(t) = V[I(t)\cos(2\pi f_c t + \varphi_c) - Q(t)\sin(2\pi f_c t + \varphi_c)] \qquad [7.60]$$

By using the result of relation [7.57], the associated analytical signal is then written:

$$s_a(t) = V\big[I(t)\exp[j(2\pi f_c t + \varphi_c)] + jQ(t)\exp[j(2\pi f_c t + \varphi_c)]\big] \qquad [7.61]$$

or:

$$s_a(t) = V[I(t) + jQ(t)]\exp[j(2\pi f_c t + \varphi_c)] \qquad [7.62]$$

By writing:

$$s_e(t) = V[I(t) + jQ(t)] = Vc(t) \qquad [7.63]$$

The signal $s_a(t)$ is then written:

$$s_a(t) = Vc(t)\exp[j(2\pi f_c t + \varphi_c)] = s_e(t)\exp[j(2\pi f_c t + \varphi_c)] \qquad [7.64]$$

The signal $s(t)$ is obtained from $s_a(t)$ by:

$$s(t) = \Re[s_a(t)] = \Re\big[s_e(t)\exp[j(2\pi f_c t + \varphi_c)]\big] \qquad [7.65]$$

Relation [7.65] shows that having knowledge of the signal $s_e(t)$ and the frequency f_c are sufficient to describe the signal $s(t)$. The signal $s_e(t)$ is the *complex envelope* of the signal $s(t)$. Its *rigorous definition* is given by:

$$s_e(t) = s_a(t)\exp[-j(2\pi f_c t + \varphi_c)] \qquad [7.66]$$

The relationship between the band-pass filter and the equivalent low-pass filter validates all these results.

7.9. Relationship between band-pass filter *H* and equivalent low-pass filter H_e

Let $s(t)$ be the modulated signal:

$$s(t) = \Re\big[s_e(t)\exp[j(2\pi f_c t + \varphi_c)]\big]$$

The output of the global band-pass filter H when its input is $s(t)$ is:

$$s_c(t) = s(t) \otimes h(t) \qquad [7.67]$$

for $h(t)$ real, the signal $s_c(t)$ is written:

$$\begin{aligned} s_c(t) &= \Re[s_e(t)\exp[j(2\pi f_c t + \varphi_c)] \otimes h(t)] \\ &= \Re\left[\exp[j(2\pi f_c t + \varphi_c)] \int_{-\infty}^{\infty} s_e(t-\tau)\exp(-j2\pi f_c \tau) \times h(\tau)d\tau\right] \end{aligned} \qquad [7.68]$$

Its complex envelope is then:

$$s_{c,e}(t) = \int_{-\infty}^{\infty} s_e(t-\tau)\exp(-j2\pi f_c\tau) \times h(\tau)d\tau \quad [7.69]$$

or by calling:

$$h_q(\tau) = h(\tau) \times \exp(-j2\pi f_c\tau) \quad [7.70]$$

$$s_{c,e}(t) = \int_{-\infty}^{\infty} s_e(t-\tau) \times h_q(\tau)d\tau = s_e(t)\otimes h_q(t) \quad [7.71]$$

Concerning the global band-pass filter gain $H(f)$, if we separate the frequency domain into two parts, that of the positive frequencies, noted $H^+(f)$ and that of the negative frequencies, noted $H^-(f)$, then the filter $H_q(f)$ is written:

$$H_q(f) = H(f)\otimes\delta(f+f_c) = [H^+(f) + H^-(f)]\otimes\delta(f+f_c) \quad [7.72]$$

From relations [7.71] and [7.72], we have:

$$S_{c,e}(f) = S_e(f)[H^+(f+f_c) + H^-(f+f_c)] \quad [7.73]$$

$S_e(f)$ is the baseband spectrum centered around zero frequency and does not cover the frequency band represented by $H^-(f+f_c)$. So, we can write that:

$$S_{c,e}(f) = S_e(f) \times H^+(f+f_c) = S_e(f) \times H_e(f) \quad [7.74]$$

Hence:

$$H_e(f) = H^+(f+f_c) \quad [7.75]$$

is the low-pass filter equivalent to the band-pass filter $H(f)$.

We summarize in Figure 7.19 all the operations, which we have justified, from a carrier digital transmission to an equivalent baseband digital transmission, thanks to relation [7.75] on the one hand, and to the definition of the complex envelope of signals on the other hand. All this confirms that the *operation of filtering carried out materially in digital transmission with carrier can be performed via simulation with the equivalent baseband digital transmission.*

Also, as $h(t)$ is any real function, then:

$$H(-f) = H^*(f) \ ; \ H(f) = H^*(-f) \tag{7.76}$$

verifies the *Hermitian symmetry*, i.e.:

$$\begin{cases} \Re\{H(-f)\} = \Re\{H(f)\} & \text{is an even function} \\ \Im\{H(-f)\} = -\Im\{H(f)\} & \text{is an odd function} \end{cases} \tag{7.77}$$

In these conditions:

$$H(f) = H^+(f) + H^-(f) = H_e(f - f_c) + H_e^*(-f - f_c) \tag{7.78}$$

Hence:

$$\begin{aligned} h(t) &= h_e(t)\exp[j(2\pi f_c t + \varphi_c)] + h_e^*(t)\exp[-j(2\pi f_c t + \varphi_c)] \\ &= 2\Re\{h_e(t)\exp[j(2\pi f_c t + \varphi_c)]\} \end{aligned} \tag{7.79}$$

By writing:

$$h_e(t) = h_{e,p}(t) + jh_{e,q}(t) \tag{7.80}$$

We then have:

$$h(t) = 2\left[h_{e,p}(t)\cos(2\pi f_c t + \varphi_c) - h_{e,q}(t)\sin(2\pi f_c t + \varphi_c)\right] \tag{7.81}$$

Moreover, we can write that:

$$\begin{aligned} s_c(t) &= \Re\left[s_{c,e}(t)\exp[j(2\pi f_c t + \varphi_c)]\right] \\ &= \frac{1}{2}\left[s_{c,e}(t)\exp[j(2\pi f_c t + \varphi_c)] + s_{c,e}^*(t)\exp[-j(2\pi f_c t + \varphi_c)]\right] \end{aligned} \tag{7.82}$$

Its amplitude spectrum is then:

$$\begin{aligned} S_c(f) &= \frac{1}{2}[S_{c,e}(f)\otimes\delta(f - f_c) + S_{c,e}^*(-f)\otimes\delta(-f - f_c)] \\ &= \frac{1}{2}[S_{c,e}(f - f_c) + S_{c,e}^*(-f - f_c)] \end{aligned} \tag{7.83}$$

But, as:

$$S_{c,e}(f) = S_e(f) \times H_e(f) \quad [7.84]$$

hence:

$$S_c(f) = \frac{1}{2}[S_e(f - f_c) \times H_e(f - f_c) + S_e^*(-f - f_c) \times H_e^*(-f - f_c)] \quad [7.85]$$

The response of the equivalent low-pass filter to the information-bearing signal $c(t)$ is:

$$s_{c,e}(t) = s_e(t) \otimes h_e(t) = V[I(t) + jQ(t)] \otimes \left[h_{e,p}(t) + jh_{e,q}(t)\right] \quad [7.86]$$

and:

$$\begin{aligned} s_{c,e}(t) &= V\left[I(t) \otimes h_{e,p}(t) - Q(t) \otimes h_{e,q}(t)\right] \\ &\quad + jV\left[I(t) \otimes h_{e,q}(t) + Q(t) \otimes h_{e,p}(t)\right] \end{aligned} \quad [7.87]$$

Figure 7.18 illustrates the construction of the complex envelope $s_{c,e}(t)$ from the information-bearing signal $c(t)$.

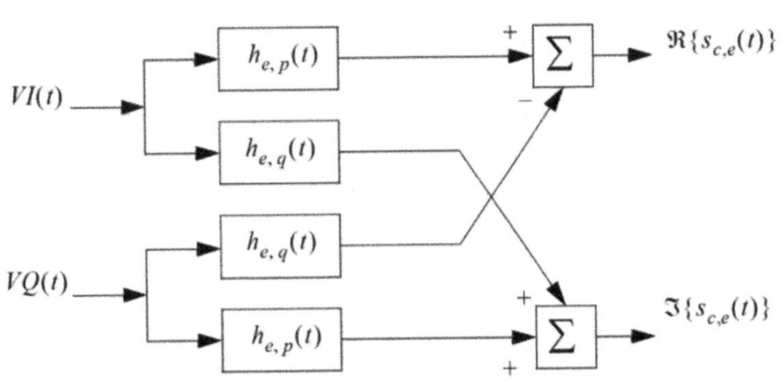

Figure 7.18. *Construction of the complex envelope $s_{c,e}(t)$ from the information-bearing signal $c(t)$*

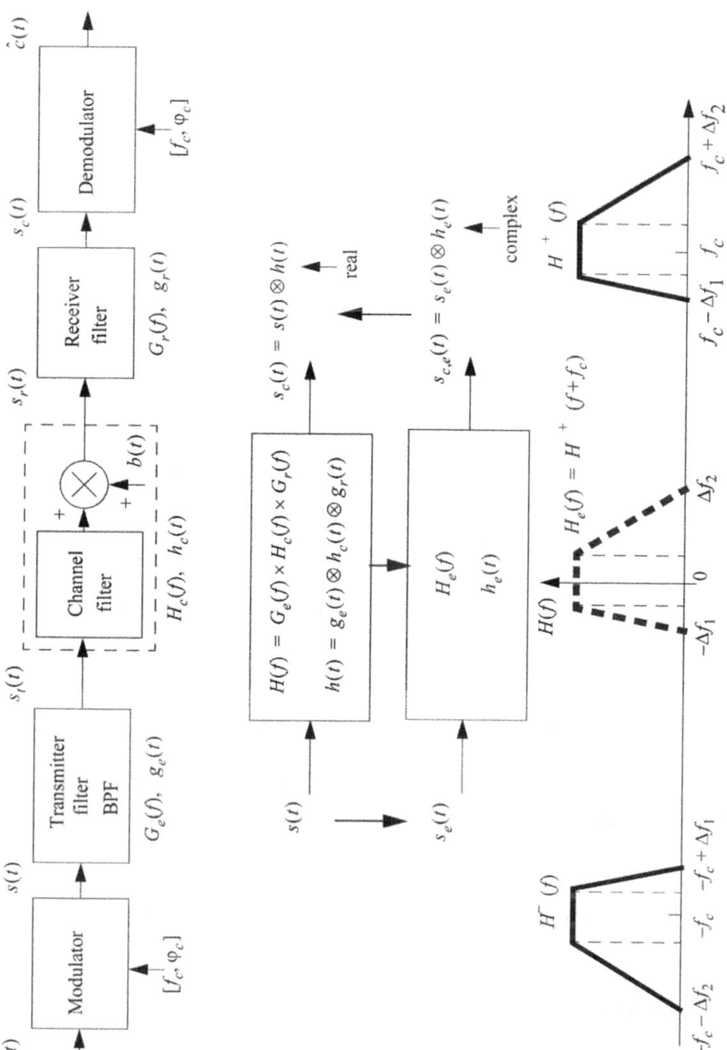

Figure 7.19. *Transition from a digital transmission with carrier modulation to an equivalent baseband digital transmission. For a color version of this figure, see www.iste.co.uk/assad/digital1.zip*

7.9.1. *Probability of errors*

The calculation of the probability of error in equivalent baseband digital transmission is performed in exactly the same way as in the case of a truly *baseband digital transmission*. All the error probability relationships that have been established in baseband transmission remain valid in equivalent baseband transmission, by replacing, in the expressions of the noise intervals and the optimal decision threshold, $p(t_0)$ with $Vp(t_0)$. For example, in the case of transmission of binary symbols $a_k = \pm 1$, relations [6.41], [6.46] and [6.59] in equivalent baseband transmission, become, respectively:

$$P_{e_{-1}} = P_{e(1/-1,m_l)} = \frac{1}{\sigma_e\sqrt{2\pi}} \int_{\mu_{0,e}+Vp(t_0)-I_{m_l}(t_0+kT)}^{\infty} \exp\left[-\frac{1}{2}\frac{b_{e,p}^2}{\sigma_e^2}\right] db_{e,p} \quad [7.88]$$

$$P_{e_1} = P_{e(-1/1,m_l)} = \frac{1}{\sigma_e\sqrt{2\pi}} \int_{-\infty}^{\mu_{0,e}-\left[Vp(t_0)+I_{m_l}(t_0+kT)\right]} \exp\left[-\frac{1}{2}\frac{b_{e,p}^2}{\sigma_e^2}\right] db_{e,p} \quad [7.89]$$

$$\mu_{0,e} = \mu_{0opt,e} = \frac{\sigma_e^2}{2Vp(t_0)}\log_e\left[\frac{p_{-1}}{p_1}\right] \quad [7.90]$$

In the following sections, we will present the different types of linear digital modulations.

7.10. M-ary Phase Shift Keying Modulation (M-PSK)

By setting $\rho_n = 1$ in the general relations [7.27] and [7.29] of joint amplitude and phase shift keying modulation, we obtain the expression of a phase shift keying digital modulation:

$$s(t) = V\sum_n x(t-nT)\cos(2\pi f_c t + \varphi_c + \psi_n) \quad [7.91]$$

and, by developing the cosine:

$$s(t) = V\sum_n \cos(\psi_n)\, x(t-nT)\cos(2\pi f_c t + \varphi_c) - V\sum_n \sin(\psi_n)\, x(t-nT)\sin(2\pi f_c t + \varphi_c) \quad [7.92]$$

where:

$$\psi_n = \frac{\pi}{M} + 2k\frac{\pi}{M} \qquad [7.93]$$

and k randomly takes the values $k = \{0,1, \ldots, M-1\}$ over the interval $nT \leq t < (n+1)T$ and $M > 2$. Furthermore, remember that, by taking:

$$a_n = \cos(\psi_n) \;;\; a_n^{'} = \sin(\psi_n)$$

the relation [7.92] can also be interpreted as a Quadrature Amplitude Modulation (QAM).

7.10.1. *Binary Phase Shift Keying (BPSK) modulation and demodulation*

7.10.1.1. *BPSK modulator*

Binary phase shift keying modulation is very simple. It is characterized by:

$$M = 2 \;;\; \psi_n = 0 \text{ and } \pi \;\leftrightarrow\; a_n = 1 \text{ and } -1 \qquad [7.94]$$

The transmission is done with only the use of the in-phase carrier. The in-quadrature carrier is not used because there are no symbols $a_n^{'}$ to transmit. The block diagram of the modulator and the corresponding constellation diagram are shown in Figure 7.20. The modulator is fully similar to that of a two-state amplitude modulation.

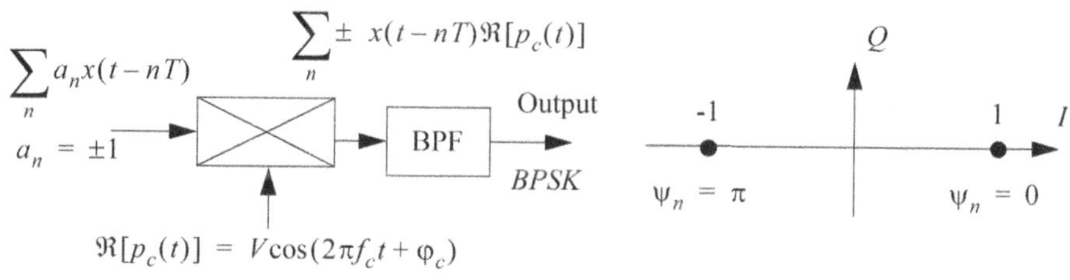

Figure 7.20. *Block diagram of the BPSK modulator and constellation diagram*

7.10.1.2. *BPSK demodulator*

The demodulation of the BPSK signal is presented in the block diagram of the BPSK demodulator, given in Figure 7.21, below.

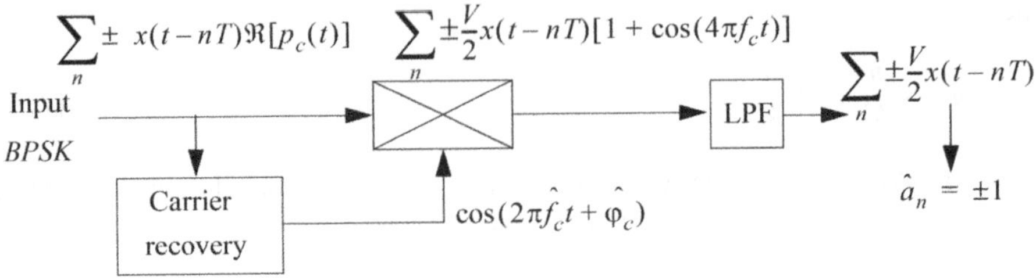

Figure 7.21. *Block diagram of the BPSK demodulator*

7.10.1.3. *Spectral occupancy of the BPSK signal*

Recall, in the general case of a M-state modulation, the relation between the bitrate, the symbol rate and the number of states:

$$\begin{cases} D(\text{bit/s}) = 1/T_b\,; & D_s(\text{symbol/s}) = 1/T \\ T = T_b \log_2 M & ;\; M = 2^k \;\; \rightarrow D_s = D/k \end{cases} \quad [7.95]$$

where k is the number of bits to group to form the M-ary symbol.

In the case of BPSK modulation:

$$M = 2 \;\rightarrow\; D_s = D \quad [7.96]$$

The spectral occupancy of the BPSK signal can be evaluated as follows. We have already seen that the frequency band occupied by the modulated signal is equal to the frequency band B of the modulating signal. The latter is maximum B_{Max}, when the binary sequence alternately changes between the 1 and 0, as shown in Figure 7.22, below. In this worst case scenario, the fundamental frequency of the binary sequence is:

$$f_a = 1/T_a = 1/2T_b = D/2 \quad [7.97]$$

and the output of the BPSK modulator is written as:

$$\begin{aligned} s(t) &= \cos(2\pi f_a t) \times V\cos(2\pi f_c t) \\ &= \frac{V}{2}\{\cos[2\pi(f_c - f_a)t] + \cos[2\pi(f_c + f_a)t]\} \end{aligned} \quad [7.98]$$

The first Nyquist criterion states that, to transmit a signal of symbol rate $D_S = 1/T$ without ISI, a minimum frequency bandwidth of the transmission channel is required:

– $f_N = 1/2T$ in the case of a baseband digital transmission, and;

– $f_N = 1/T$ in the case of a digital transmission with carrier modulation.

Figure 7.23 illustrates this.

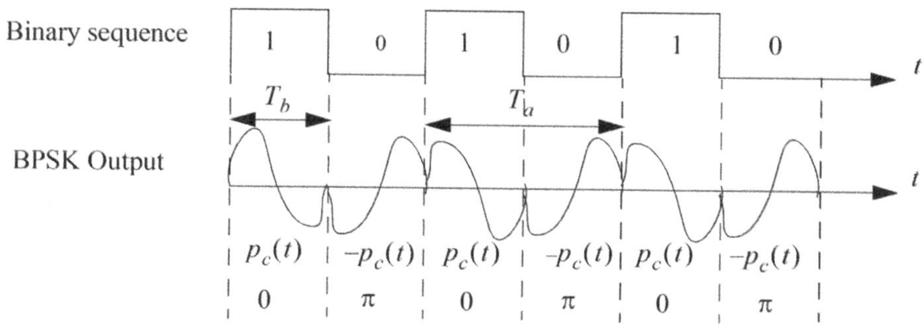

Figure 7.22. *Output of the BPSK modulator for a binary sequence that changes alternately between the values 1 and 0*

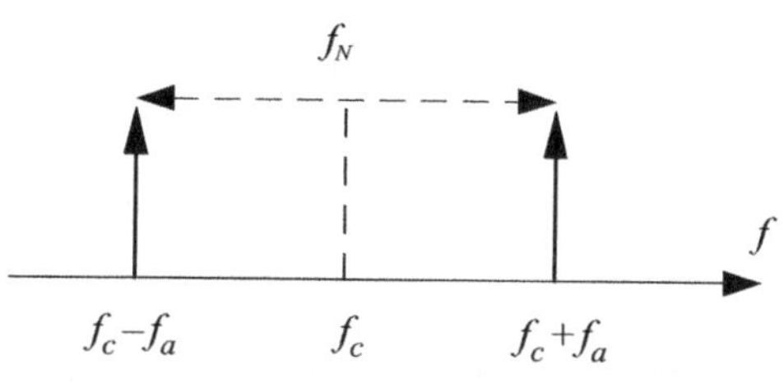

Figure 7.23. *Minimum frequency bandwidth required for a channel to transmit witout ISI a BPSK signal*

Indeed, the Nyquist frequency, according to Figure 7.23, is:

$$f_N = [f_c + f_a] - [f_c - f_a] = 2f_a = D = D_s \quad [7.99]$$

with:

– $f_c - f_a$ as the lower value of the frequency band (mininum lower side frequency: LSF);

– $f_c + f_a$ as the upper value of the frequency band (maximun upper side frequency: USF).

EXAMPLE.– Let us take a BPSK modulator with $f_c = 70\text{ MHz}$ and $D = 10\text{ Mbit/s}$.

Calculate:

1) The fundamental frequency f_a and the symbol rate D_s;

2) The lower value of the frequency band (LSF);

3) The upper value of the frequency band (USF);

4) The Nyquist frequency.

SOLUTION.–

1) The fundamental frequency is $f_a = D/2 = 5\text{ MHz}$ and the symbol rate is $D_s = D = 10\text{ Msymbol/s}$.

2) The lower value of the frequency band (LSF) is $f_c - f_a = 70 - 5 = 65\text{ MHz}$.

3) The upper value of the frequency band (USF) is $f_c + f_a = 70 + 5 = 75\text{ MHz}$.

4) The Nyquist frequency is $f_N = 2f_a = 10\text{ MHz}$.

7.10.2. *Quaternary Phase Shift Keying (QPSK) modulation and demodulation*

The modulated signal is that given by the relation [7.92], which we recall below:

$$s(t) = V\sum_n \cos(\psi_n)\,x(t - nT)\cos(2\pi f_c t + \varphi_c)$$

$$-V\sum_n \sin(\psi_n)\,x(t - nT)\sin(2\pi f_c t + \varphi_c)$$

with:

$$M = 4 \ ;$$

$$\psi_n = \frac{\pi}{4} + k\frac{\pi}{2} \ ; \qquad [7.100]$$

$$k = \{0, 1, 2, 3\}$$

By writing:

$$a_n = \cos(\psi_n) \ ; \ a'_n = \sin(\psi_n)$$

we get the relation [7.31]:

$$s(t) = V\sum_n a_n\, x(t - nT)\cos(2\pi f_c t + \varphi_c) - V\sum_n a'_n\, x(t - nT)\sin(2\pi f_c t + \varphi_c)$$

It is this relationship that is implanted by the QPSK modulator shown in Figure 7.24, below.

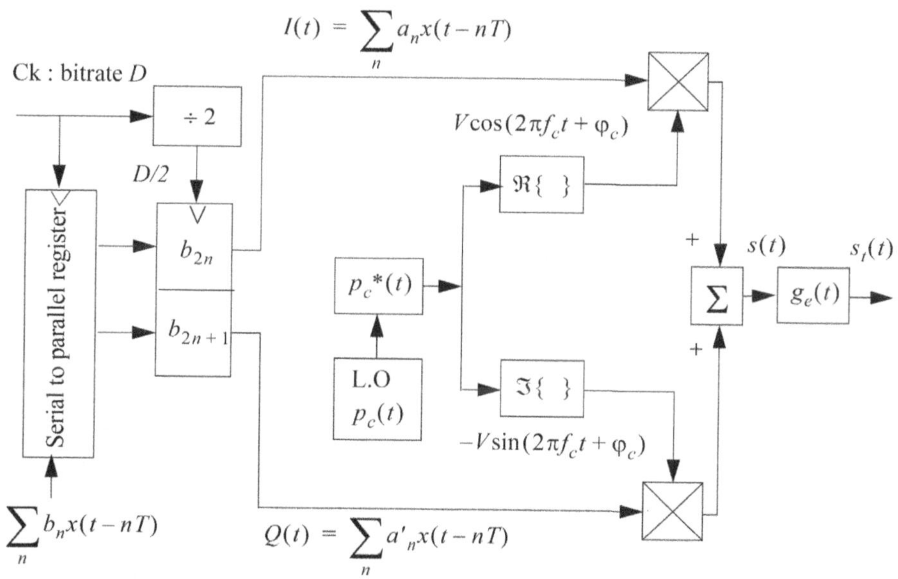

Figure 7.24. *Block diagram of the QPSK modulator*

It is designed according to the general model of the linear modulator in Figure 7.12, and operates as follows.

The binary signal $\sum_n b_n \, x(t-nT)$ enters a serial register, formed of two D flip-flops clocked at the frequency $1/T_b$. This allows us to split it into two baseband signals $I(t) = \sum_n a_n \, x(t-nT)$ and $Q(t) = \sum_n a_n^{'} \, x(t-nT)$ each clocked at the frequency $1/2T_b$, with: $a_n = [1 - 2b_{2n}] = \pm 1$ and $a_n^{'} = [1 - 2b_{2n+1}] = \pm 1$. The signal $I(t)$ then modulates the in-phase carrier produced by a local oscillator (OL) in amplitude, and the signal $Q(t)$ modulates the in-quadrature carrier produced by the local oscillator (OL) in amplitude. The two resulting components are summed to form the modulated signal $s(t)$. Finally, it is filtered by the band-pass emission filter $g_e(t)$ centered around the frequency f_c to limit the spectral range of the signal $s(t)$ (of the form $\sin(\pi f t)/(\pi f t)$) to the band B and to ensure optimal filtering between transmission and reception. The signal effectively transmitted is $s_t(t)$.

To form the coding table and the vector diagram, we rewrite the modulated signal $s(t)$ according to [7.91]:

$$s(t) = V \sum_n x(t-nT) \cos\left(2\pi f_c t + \varphi_c + \frac{\pi}{4} + k\frac{\pi}{2}\right) \quad k = \{0,1,2,3\} \qquad [7.101]$$

Hence, the coding table and the vector diagram given in Figure 7.25.

$b_{2n+1} \, b_{2n}$	$a'_n \quad a_n$	ψ_n
0 0	1, 1	45°
0 1	1, –1	135°
1 1	–1, –1	225°
1 0	–1, 1	315°

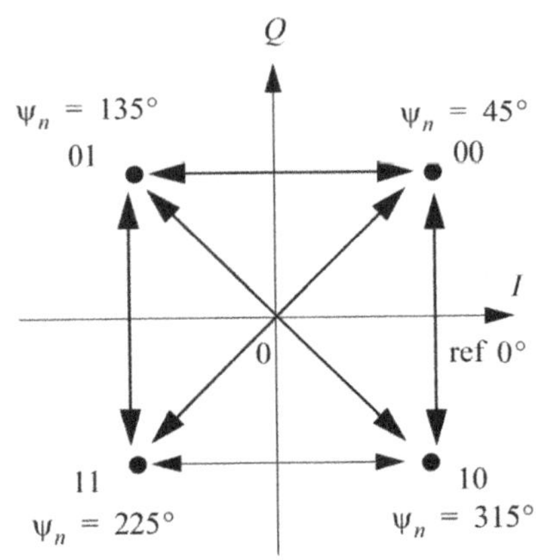

Figure 7.25. *Coding table and vectorial diagram of QPSK modulation*

7.10.2.1. *Spectral occupancy of the QPSK signal*

In the case of the QPSK modulated signal, one has:

$$M = 4 \;\rightarrow\; D_s = D_I = D_Q = D/2 \qquad [7.102]$$

Following the same reasoning as in the case of the BPSK modulation, the freqency band occupied by the signal is maximum B_{Max} when the bit sequence on the I and Q channels alternately changes between the binary values 1 and 0, as shown in Figure 7.26.

In this case (see Figure 7.26), the largest fundamental frequency is:

$$f_a = 1/T_a = 1/4T_b = D/4 \qquad [7.103]$$

and the Nyquist frequency is then:

$$f_N = 2f_a = D/2 = D_s \qquad [7.104]$$

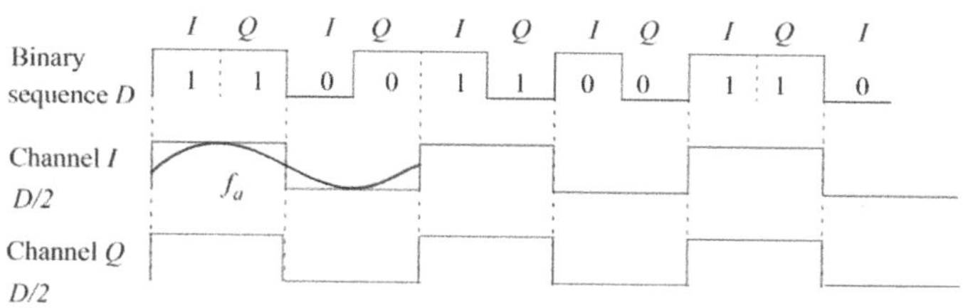

Figure 7.26. *Channels $I(t)$ and $Q(t)$ for a binary sequence that changes alternately between the binary values 1 and 0*

EXAMPLE.– Let us take a QPSK modulator, with $f_c = 70$ MHz and $D = 10\,\text{Mbit/s}$.

1) Calculate the symbol rate D_s, the fundamental frequency f_a, and the Nyquist frequency f_N.

2) Draw the spectrum of the modulated signal $s(t)$ and compare the results obtained with the BPSK modulator.

SOLUTION.–

1) The symbol rate is: $D_s = D/2 = 5\,\text{Msymbol/s}$.

The fundamental frequency is $f_a = D/4 = 2.5$ MHz.

The Nyquist frequency is $f_N = 2f_a = 5$ MHz.

2) In this case, the signal $s(t)$ is written:

$$s(t) = \frac{V}{2}\{\cos[2\pi(67.5 \times 10^6)t] + \cos[2\pi(72.5 \times 10^6)t]\}$$

and its spectrum is then:

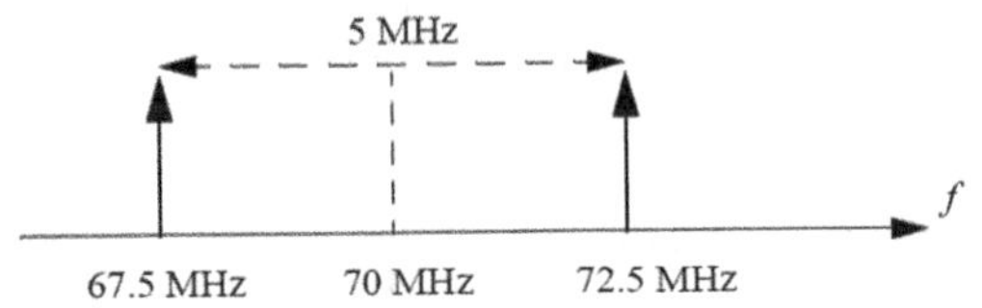

For the same bitrate D, the minimum required frequency band of the transmission channel and the symbol rate are, respectively:

$$f_{N(QPSK)} = \frac{1}{2} f_{N(BPSK)}$$

$$D_{s(QPSK)} = \frac{1}{2} D_{s(BPSK)}$$

The QPSK receiver is designed as shown in the block diagram in Figure 7.27.

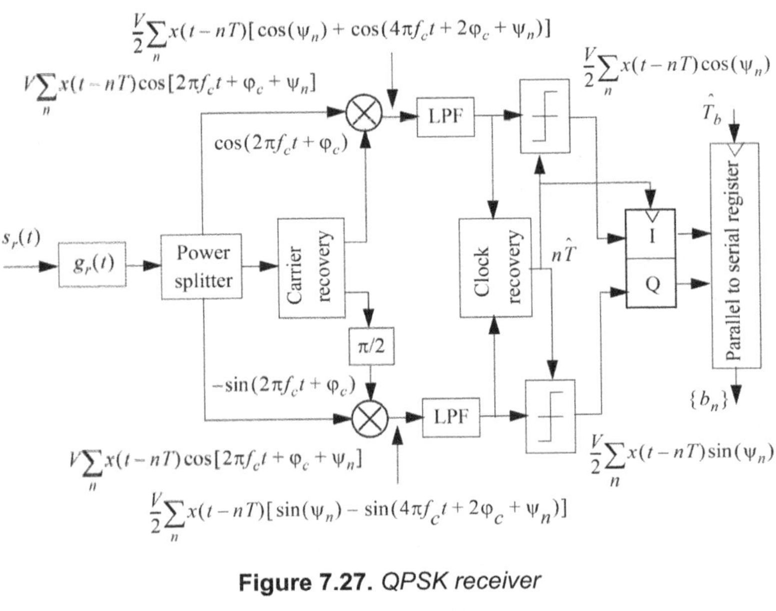

Figure 7.27. *QPSK receiver*

The received signal $s_r(t)$ is first filtered by the reception filter $g_r(t)$ which also limits the reception noise power.

Then, after demodulation by the in-phase carrier, and, in parallel, by the in-quadrature carrier, low-pass filters (LPF) cancel the high frequency part of each component.

Both decision systems then generate the symbol sequences $\{\hat{a}_n\}$ and $\{\hat{a}'_n\}$ and the transcoding unit provides the restored bit sequence $\{\hat{b}_n\}$.

7.10.3. *Differential M-PSK receiver*

The carrier recovery unit of the M-PSK receivers contains a circuit that raises, at power M, the received signal. This causes a *loss of phase information (phase anbiguity)*.

To solve this problem, *non-absolute coding of information by relative phase coding (differential coding)* is used.

7.10.3.1. *Differential BPSK coding and decoding*

In Figure 7.28, we give the coding and decoding table of the differential BPSK code, as well as the block diagram of the implementation of the differential coding and decoding circuits.

ψ_{n-1}	I_{n-1}	b_n	$\Delta\psi_n$	ψ_n	I_n
0°	0	0	0°	0°	0
0°	0	1	180°	180°	1
180°	1	0	0°	180°	1
180°	1	1	180°	0°	0

Figure 7.28(a). *Differential BPSK coding and decoding table*

Coding law : $I_n = b_n\overline{I_{n-1}} + \overline{b_n}I_{n-1} = b_n \oplus I_{n-1}$

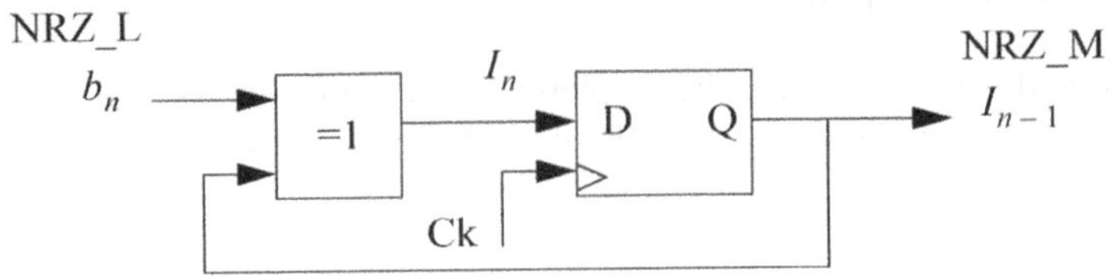

Decoding law : $b_n = \overline{I_{n-1}}I_n + I_{n-1}\overline{I_n} = I_{n-1} \oplus I_n$

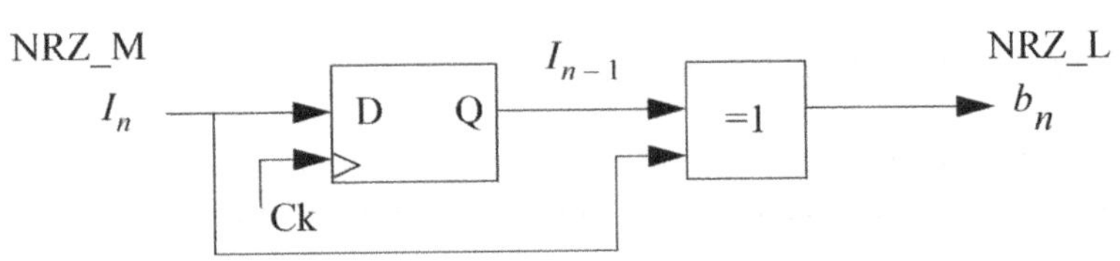

Figure 7.28(b). *Associated circuits (following)*

The coding and decoding table is very explicit. It shows that the coding of a bit I_n at time n depends on the coding of the bit I_{n-1} at time $n-1$ and on the value of the binary symbol b_n. Depending on the value of the latter, the phase jump $\Delta\psi_n$ takes the value "0°" or "180°", and the phase ψ_n is then given by $\psi_n = \psi_{n-1} + \Delta\psi_n$.

7.10.3.2. *Differential QPSK coding and decoding*

Differential QPSK coding and decoding are constructed in the same way as differential BPSK coding and decoding. In Figure 7.29, we give the differential QPSK coding and decoding table, as well as the two Karnaugh tables for the simplification of the equations Q_n and I_n.

The differential coding and decoding table is fully explained. It shows that the coding of a couple of binary symbols Q_nI_n at time n depends on the coding of the pair of bits $Q_{n-1}I_{n-1}$ at time $n-1$ and on the value of the pair of binary symbols $b_{2n+1}b_{2n}$ to be coded. Depending on the value of the pair of binary symbols, the phase jump $\Delta\psi_n$ takes the value "0°", "90°", "180°" or "270°" and the phase ψ_n is given by the relation $\psi_n = \psi_{n-1} + \Delta\psi_n$. In Figure 7.30, we give the simplified relations of the encoder as well as its electric implementation. In Figure 7.31, we give the two simplified equations, by Karnaugh tables, of the decoder and in Figure 7.32 we give its electric implementation.

ψ_{n-1}	Q_{n-1}	I_{n-1}	b_{2n+1}	b_{2n}	$\Delta\psi_n$	ψ_n	Q_n	I_n
45°	0	0	0	0	0°	45°	0	0
			0	1	90°	135°	0	1
			1	1	180°	225°	1	1
			1	0	270°	315°	1	0
135°	0	1	0	0	0°	135°	0	1
			0	1	90°	225°	1	1
			1	1	180°	315°	1	0
			1	0	270°	45°	0	0
225°	1	1	0	0	0°	225°	1	1
			0	1	90°	315°	1	0
			1	1	180°	45°	0	0
			1	0	270°	135°	0	1
315°	1	0	0	0	0°	315°	1	0
			0	1	90°	45°	0	0
			1	1	180°	135°	0	1
			1	0	270°	225°	1	1

Q_n:

$Q_{n-1}\ I_{n-1}$ \ $b_{2n+1}\ b_{2n}$	00	01	11	10
00			1	1
01		1	1	
11	1	1		
10	1			1

I_n:

$Q_{n-1}\ I_{n-1}$ \ $b_{2n+1}\ b_{2n}$	00	01	11	10
00		1	1	
01	1	1		
11	1			1
10			1	1

Figure 7.29. *Differential QPSK coding and decoding table and Karnaugh tables*

$$Q_n = \overline{Q_{n-1}}\,\overline{I_{n-1}}\,b_{2n+1} + \overline{Q_{n-1}}\,I_{n-1}\,b_{2n} + Q_{n-1}\,\overline{I_{n-1}}\,\overline{b_{2n}} + Q_{n-1}\,I_{n-1}\,\overline{b_{2n+1}}$$

$$I_n = \overline{Q_{n-1}}\,\overline{I_{n-1}}\,b_{2n} + \overline{Q_{n-1}}\,I_{n-1}\,\overline{b_{2n+1}} + Q_{n-1}\,\overline{I_{n-1}}\,b_{2n+1} + Q_{n-1}\,I_{n-1}\,\overline{b_{2n}}$$

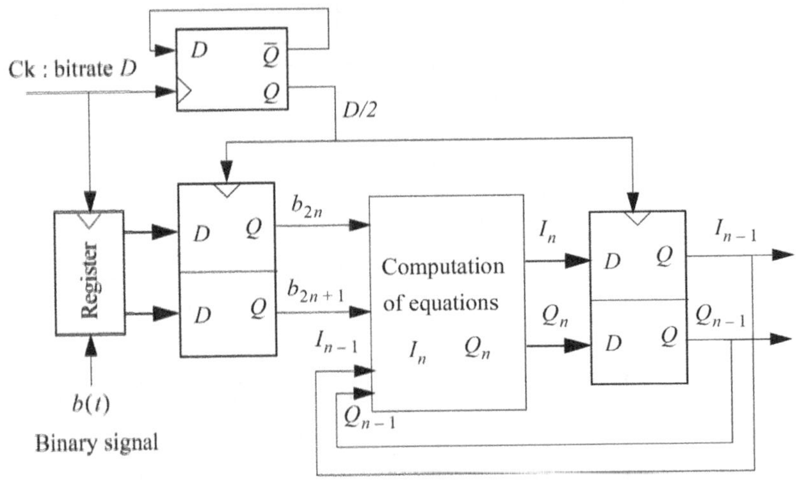

Figure 7.30. *Simplified equations of the differential QPSK encoder and its implementation scheme*

7.10.3.3. *Differential QPSK decoder*

b_{2n+1}:

$Q_n I_n$ \ $Q_{n-1} I_{n-1}$	00	01	11	10
00		1	1	
01			1	1
11	1			1
10	1	1		

b_{2n}:

$Q_n I_n$ \ $Q_{n-1} I_{n-1}$	00	01	11	10
00			1	1
01	1			1
11	1	1		
10		1	1	

$$b_{2n+1} = \overline{Q}_n\overline{I}_n I_{n-1} + \overline{Q}_n I_n Q_{n-1} + Q_n\overline{I}_n\overline{Q_{n-1}} + Q_n I_n \overline{I_{n-1}}$$

$$b_{2n} = \overline{Q}_n\overline{I}_n Q_{n-1} + \overline{Q}_n I_n \overline{I_{n-1}} + Q_n\overline{I}_n I_{n-1} + Q_n I_n \overline{Q_{n-1}}$$

Figure 7.31. *Simplified equations of the differential QPSK decoder by Karnaugh tables*

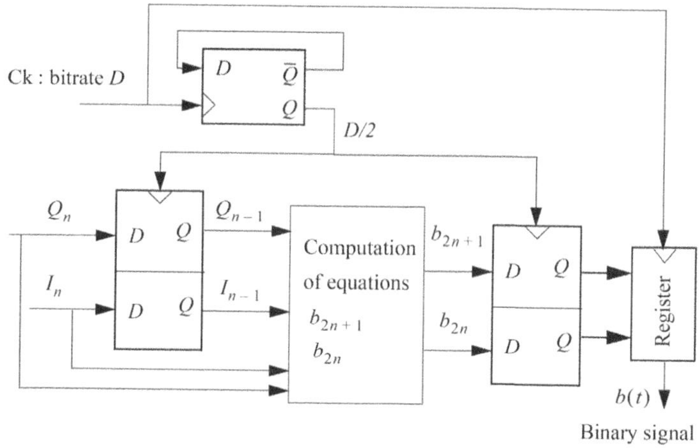

Figure 7.32. *Implementation scheme of the differential QPSK decoder*

7.10.4. *Offset Quaternary Phase Shift Keying (OQPSK)*

In QPSK modulation, and for most modulations, the phase jump $\Delta\psi_n$ between phase states is variable. We can therefore have a phase jump $\Delta\psi_n = \pm\pi$ for the transition from the state $s[nT]$ to the state $s[(n+1)T]$ (see Figure 7.25). This means that the amplitude of the received signal passes through a zero value at a given instant. This therefore brings about degradation and reduction of the complex envelope $s_e(t)$ and, consequently, the possible loss of synchronization in reception.

To overcome this problem, one of the solutions is to use *time offset phase modulations* to avoid phase jump $\Delta\psi_n = \pm\pi$.

The QPSK modulated signal with time offset (OQPSK) can then be written:

$$s(t) = V\sum_n b_n x(t - nT)\cos\left(2\pi f_c t + \varphi_c + n\frac{\pi}{2}\right) \qquad [7.105]$$

where b_n is the binary sequence and the phase jump between two constellations is fixed and equal to:

$$\Delta\psi_n = \pm\frac{\pi}{2} \qquad [7.106]$$

The complex envelope of the modulated signal will never go through the zero value (see Figure 7.33).

The relation [7.105] can also be written:

$$s(t) = V\sum_{n} a_n x(t - nT)\cos(2\pi f_c t + \varphi_c) - V\sum_{n} a'_n x \left(t - \left[n + \frac{1}{2}\right]T\right)\sin(2\pi f_c t + \varphi_c) \quad [7.107]$$

with:

$$a_n = (-1)^n[2b_{2n} - 1] \;\; ; \;\; a'_n = (-1)^n[2b_{2n+1} - 1] \quad [7.108]$$

If, in the relation [7.107], we replace:

$$x(t) = Rect_T(t) \text{ with } x(t) = Rect_T(t)\sin\left[\frac{\pi t}{T}\right] \quad [7.109]$$

we get the minimum phase shift keying modulation (MSK).

In Figure 7.33, we present the binary and modulating signals, the phases and the constellation diagram of time-offset QPSK modulation (OQPSK) (see also Figure 7.25).

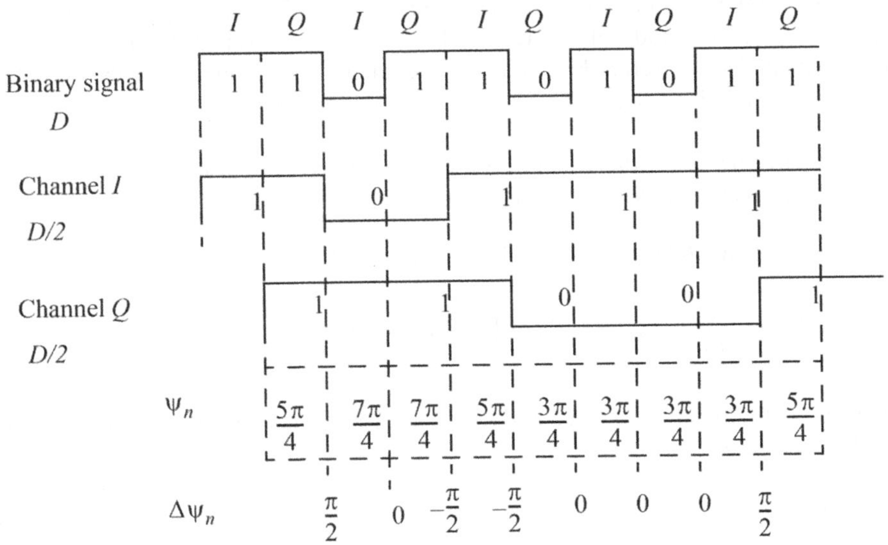

Figure 7.33(a). *Binary signal, modulating signal and phases of the OQPSK modulation*

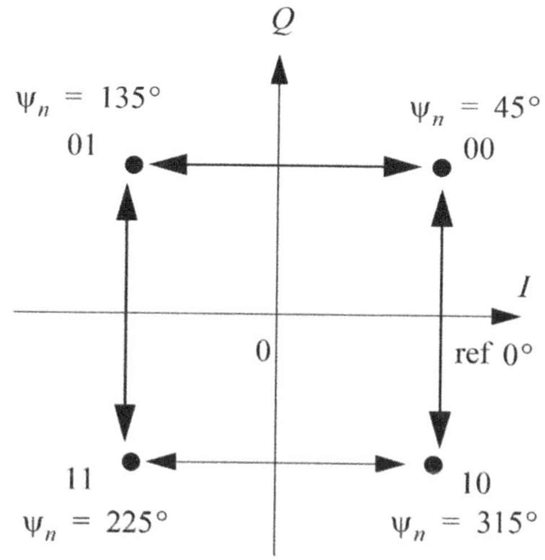

Figure 7.33(b). *Constellation diagram of the OQPSK modulation (following)*

7.10.4.1. *OQPSK modulator and demodulator*

The implementation schemes of the OQPSK modulator and demodulator are virtually identical to the QPSK modulator and demodulator schemes.

Indeed, the differences are:

– at the modulator level, after differential coding, the signal on the channel Q is delayed by $T/2$ (see Figure 7.33);

– at the demodulator level, before differential decoding, the signal on the channel I is delayed by $T/2$.

7.10.4.2. *Spectral occupancy of the OQPSK signal*

In the case of the OQPSK modulated signal:

$$M = 4 \rightarrow D_S = D_I = D_Q = D \quad [7.110]$$

Indeed, because of the time shift $T/2$ of the signal on channel Q, the speed in the phase change is twice that of the data of the channel I or the channel Q (see Figure 7.33).

Therefore, for the same bitrate D, the minimum required bandwidth of the transmission channel is:

$$f_{N(OQPSK)} = 2f_{N(QPSK)} \qquad [7.111]$$

and the symbol rate is:

$$D_{s(OQPSK)} = 2D_{s(QPSK)} \qquad [7.112]$$

7.11. M-ary Quadrature Amplitude Modulation (M-QAM)

The M-QAM modulation is identical to the M-APSK (amplitude and phase shift keying) modulation.

Therefore, the general structure of the linear modulator of Figure 7.12 and the principle of the QPSK modulator scheme of Figure 7.24 apply.

The advantage of this modulation is as follows. When M is large, the use of two channels I and Q then allows a homogeneous distribution of M states in the plane (I, Q), thus allowing many possibilities. We can have the following distributions:

– *16-QAM modulation*: four amplitudes for I and, simultaneously, four amplitudes for Q:

$$M = 16 = 2^4 \leftrightarrow \{a_n, a'_n\} = \{2 \text{ bits}, 2 \text{ bits}\}$$

This is mainly used with a bitrate of 9,600 bit/s on a telephone line, and 140 Mbit/s on the radio link. It will be studied in detail in the following section. The increase in the number of states generates an increase in the bitrate. This explains the very high speed modulations used in the LTE standard, such as:

– *64-QAM modulation*:

$$M = 64 = 2^6 \leftrightarrow \{a_n, a'_n\} = \{3 \text{ bits}, 3 \text{ bits}\}$$

– *256 QAM modulation*:

$$M = 256 = 2^8 \leftrightarrow \{a_n, a'_n\} = \{4 \text{ bits}, 4 \text{ bits}\}$$

7.12. Detailed presentation of 16-QAM modulation and demodulation

In this section we present the 16-QAM modulation and demodulation. For that, let's first remember the different relationships used.

The modulated signal is written:

$$s(t) = V\sum_{n} a_n\, x(t-nT)\cos(2\pi f_c t + \varphi_c)$$
$$-V\sum_{n} a'_n\, x(t-nT)\sin(2\pi f_c t + \varphi_c) \quad [7.113]$$

with:

$$\{a_n\} \text{ and } \{a'_n\} \in \{\pm 1, \pm 3\} \rightarrow \text{equidistance between adjacent states} \quad [7.114]$$

and:

$$\rho_n = \left[a_n^2 + a_n'^2\right]^{1/2}$$
$$\psi_n = \tan^{-1}\left[\frac{a'_n}{a_n}\right] \quad [7.115]$$

In Figure 7.34, we give the Gray coding tables of symbol a'_n and a_n as a function of the bits $(R_n\, J_n)$ and $(Q_n\, I_n)$, respectively, then the complete Gray's table for the symbol pair $(a'_n\, a_n)$ and their module and phase $(\rho_n\, \psi_n)$ as a function of the bits $(R_n\, J_n\, Q_n\, I_n)$.

Figure 7.35 shows the constellation diagram corresponding to the complete Gray's coding table above. We clearly notice the equidistance between adjacent states.

Figure 7.36 shows the block diagram of the 16-QAM modulator with differential coding in accordance with the coding table and the constellation diagram of Figures 7.34 and 7.35, respectively. The DAC (Digital to Analog Converter) block provides the desired amplitudes of the symbols a_n and a'_n.

The advantage of this implementation lies in the Gray's code, which practically reduces the value of error probability on the symbols to a value very close to the probability of bit error.

R_n J_n	a'_n
0 0	1
0 1	3
1 1	–3
1 0	–1

Q_n I_n	a_n
0 0	1
0 1	3
1 1	–3
1 0	–1

R_n	J_n	Q_n	I_n	ρ_n	ψ_n	a'_n	a_n
0	0	0	0	$\sqrt{2}$	45°	1	1
		0	1	$\sqrt{10}$	18°43	1	3
		1	1	$\sqrt{10}$	–18°43	1	–3
		1	0	$\sqrt{2}$	–45°	1	–1
0	1	1	0	$\sqrt{10}$	–71°56	3	–1
		1	1	$\sqrt{18}$	–45°	3	–3
		0	1	$\sqrt{18}$	45°	3	3
		0	0	$\sqrt{10}$	71°56	3	1
1	1	0	0	$\sqrt{10}$	–71°56	–3	1
		0	1	$\sqrt{18}$	–45°	–3	3
		1	1	$\sqrt{18}$	45°	–3	–3
		1	0	$\sqrt{10}$	71°56	–3	–1
1	0	1	0	$\sqrt{2}$	45°	–1	–1
		1	1	$\sqrt{10}$	18°43	–1	–3
		0	1	$\sqrt{10}$	–18°43	–1	3
		0	0	$\sqrt{2}$	–45°	–1	1

Figure 7.34. *Gray coding tables of the 16-QAM modulation*

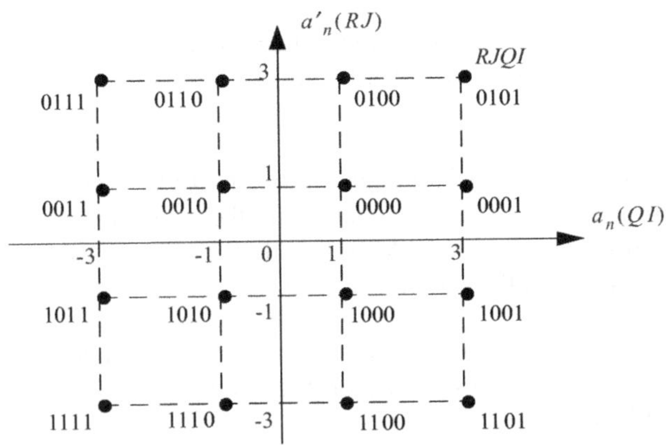

Figure 7.35. *Constellation diagram of the 16-QAM modulation*

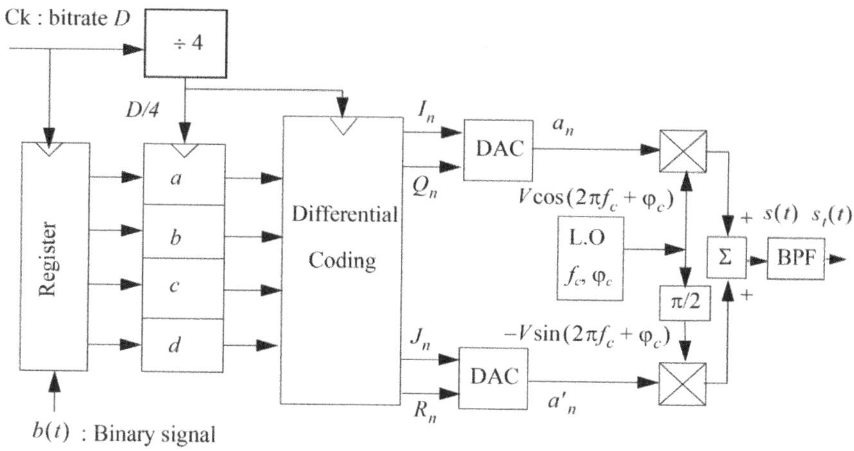

Figure 7.36. *Block diagram of the 16-QAM modulator with differential coding based on the Gray's code*

A second, simpler implementation solution is achieved by the use of two separately controlled QPSK modulators, as shown in the block diagram of Figure 7.37. However, the coding performed is different from that of Gray.

7.12.1. *Spectral occupancy of the 16-QAM modulated signal*

In the case of the 16-QAM modulation, we have:

$$M = 16 \rightarrow D_S = D_R = D_J = D_Q = D_I = D/4 \qquad [7.116]$$

The frequency band occupied by the signal is maximum B_{Max}, when the binary sequence on the R, J, Q, and I channels changes alternately between the 1 and 0 binary values (see Figure 7.38).

The highest fundamental frequency is equal to:

$$f_a = 1/T_a = 1/8T_b = D/8 \qquad [7.117]$$

and the Nyquist frequency is then:

$$f_N = 2f_a = D/4 = D_S \qquad [7.118]$$

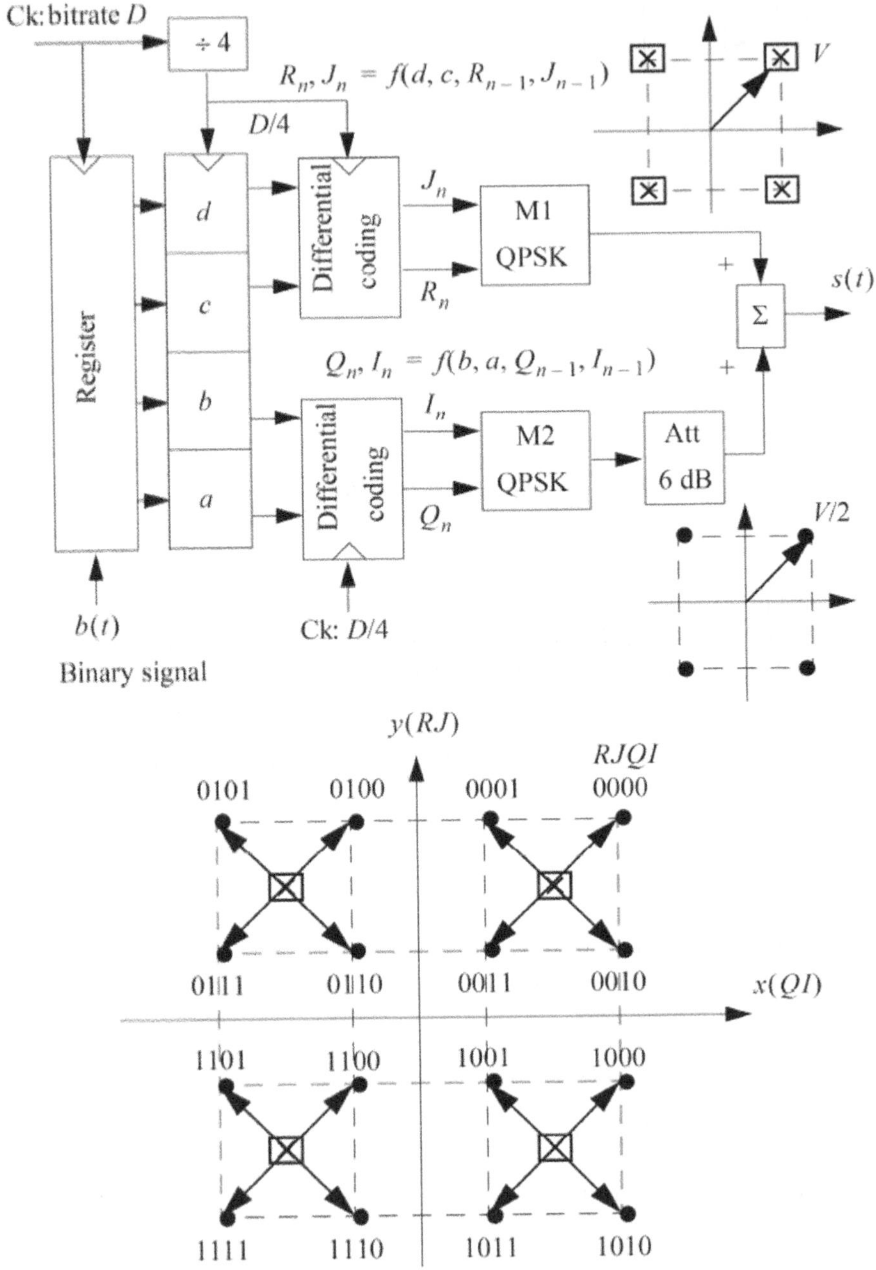

Figure 7.37. *16-QAM modulator based on the use of two separately controlled QPSK modulators*

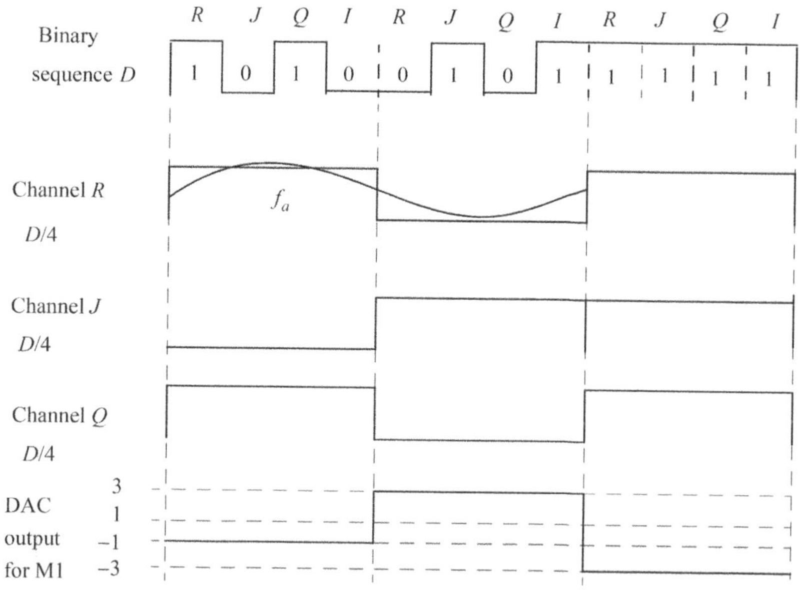

Figure 7.38. *Different channels for a binary sequence that changes alternately between binary values 1 and 0 and DAC output for M1*

EXAMPLE.– Let us take a 16-QAM modulator, with $f_c = 70\text{ MHz}$ and $D = 10\text{ Mbit/s}$.

1) Calculate the symbol rate D_s, the fundamental frequency f_a, and the Nyquist frequency f_N.

2) Display the spectrum of the modulated signal $s(t)$.

3) Compare the results obtained with modulators QPSK and BPSK.

SOLUTION.–

1) The symbol rate is $D_s = D/4 = 2.5\text{ Msymbol/s}$.

The fundamental frequency is $f_a = D/8 = 1.25\text{ MHz}$.

The Nyquist frequency is $f_N = 2f_a = 2.5\text{ MHz}$.

2) The signal $s(t)$ and its spectrum are, in this case:

$$s(t) = \frac{V}{2}\{\cos[2\pi(f_c - f_a)t] + \cos[2\pi(f_c + f_a)t]\}$$

$$s(t) = \frac{V}{2}\{\cos[2\pi(68.75 \times 10^6)t] + \cos[2\pi(71.25 \times 10^6)t]\}$$

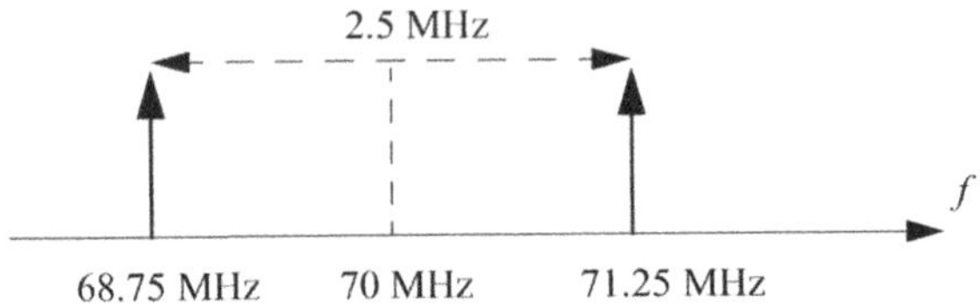

3) For the same bitrate D, the minimum required bandwidth of the channel and the symbol rate are:

$$f_{N(16-QAM)} = \frac{1}{2}f_{N(QPSK)} = \frac{1}{4}f_{N(BPSK)}$$

$$D_{s(16-QAM)} = \frac{1}{2}D_{s(QPSK)} = \frac{1}{4}D_{s(BPSK)}$$

7.13. Amplitude and Phase Shift Keying Modulation (APSK)

In this section we present the modulation CIR (4, 4, 4, 4).

7.13.1. *CIR (4, 4, 4, 4) modulation: 4 amplitudes, 4 phases*

This is a combined modulation of particular amplitude and phase, defined by Figure 7.39, for a bitrate of 9,600 bit/s, on a telephone line. The constellation diagram is constructed as follows. The amplitude ρ_n is determined by the bit R and by the absolute phase ψ_n as shown in the table at the top of Figure 7.39. The other three bits $J\ Q\ I$ are used to code the phase jump $\Delta\psi_n$ as indicated by the table at the bottom of Figure 7.39. The CIR (4, 4, 4, 4) modulator and demodulator are more complex than those of 16-QAM. However, the CIR (4, 4, 4, 4) modulation is more robust against the non-linearities of the power amplifiers than the 16-QAM modulation.

ψ_n	R	ρ_n
0°, 90°, 180°, 270°	0	3
	1	5
45°, 135°, 225°, 315°	0	$\sqrt{2}$
	1	$3\sqrt{2}$

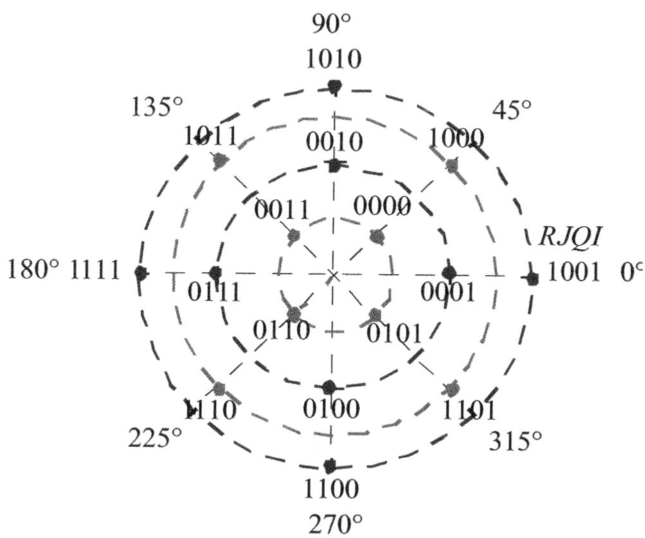

JQI	001	000	010	011	111	110	100	101
$\Delta\psi_n$	0°	45°	90°	135°	180°	225°	270°	315°

Figure 7.39. *CIR (4, 4, 4, 4) modulation. For a color version of this figure, see www.iste.co.uk/assad/digital1.zip*

7.14. Detailed presentation of the 8-PSK modulation and demoludation

The 8-PSK modulator and demodulator, apart the CIR (4, 4, 4, 4) modulation, are a little more complicated to design and implement, compared to those we have already presented.

In its general form, the modulated signal is written:

$$s(t) = V \sum_n x(t - nT) \cos(2\pi f_c t + \varphi_c + \psi_n) \qquad [7.119]$$

with:

$$M = 8 \; ; \; \psi_n = \frac{\pi}{8} + k\frac{\pi}{4} \text{ and } k = \{0, 1, \cdots, 7\} \qquad [7.120]$$

The value of the phase ψ_n is dependent on the value of the three-bit sequence $(Q_n \, I_n \, K_n)$, over the time interval $nT \leq t < (n+1)T$. Over this time interval, $s(t)$ is written:

$$\begin{aligned} s(t) &= V \sum_n a_n \, x(t - nT) \cos(2\pi f_c t + \varphi_c) \\ &-V \sum_n a'_n \, x(t - nT) \sin(2\pi f_c t + \varphi_c) \end{aligned} \qquad [7.121]$$

with:

$$a_n = \cos\left(\frac{\pi}{8} + k\frac{\pi}{4}\right) \; ; \; a'_n = \sin\left(\frac{\pi}{8} + k\frac{\pi}{4}\right) \qquad [7.122]$$

This explains the complete coding table, using the Gray code, at the top of Figure 7.40. Since the number of bits required for 8-phase coding is odd, we cannot directly use the general structure of the linear modulator of Figure 7.12. To design and implement the 8-PSK modulator (and the demodulator), according to the general model of the Figure 7.12, we take the 4 states preceded by an asterisk in the complete coding table and we complement only their bit K to find the other four states of this complete coding table (the coding table at the bottom left of Figure 7.40). On the one hand, this makes it possible to build the coding table of the symbols a'_n and a_n as a function of the bits $(Q_n \, K_n)$ and $(I_n \, \overline{K}_n)$ respectively (bottom and right table of Figure 7.40), and, on the other hand, to be able to construct the 8-PSK modulator according to the general theoretical model.

$Q_nI_nK_n$	ψ_n		a'_n	a_n
* 0 0 0	$\pi/8$	22°5	0.38	0.92
0 0 1	$3\pi/8$	67°5	0.92	0.38
* 0 1 1	$5\pi/8$	112°5	0.92	–0.38
0 1 0	$7\pi/8$	157°5	0.38	–0.92
* 1 1 0	$9\pi/8$	202°5	–0.38	–0.92
1 1 1	$11\pi/8$	247°5	–0.92	–0.38
* 1 0 1	$13\pi/8$	292°5	–0.92	0.38
1 0 0	$15\pi/8$	337°5	–0.38	0.92

$Q_n I_n K_n$	$Q_nI_n\overline{K_n}$
* 0 0 0	0 0 1
* 0 1 1	0 1 0
* 1 1 0	1 1 1
* 1 0 1	1 0 0

Q_nK_n	a'_n	a_n	$I_n\overline{K_n}$
0 0	0.38	0.92	0 1
0 1	0.92	- 0.38	1 0
1 0	- 0.38	- 0.92	1 1
1 1	- 0.92	0.38	0 0

Figure 7.40. *8-PSK modulator coding tables*

In Figure 7.41, we present the constellation diagram of the 8-PSK modulation.

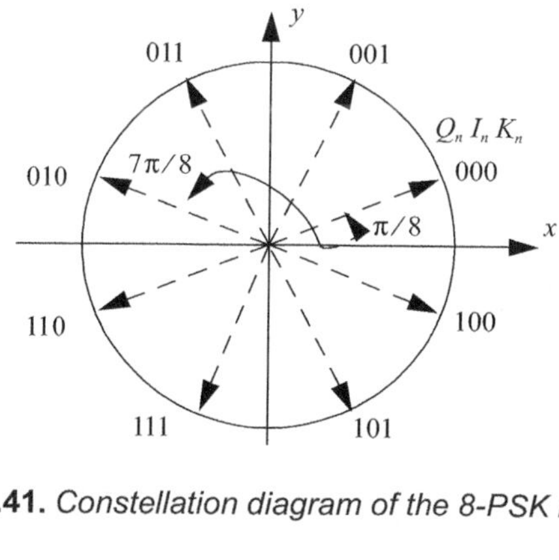

Figure 7.41. *Constellation diagram of the 8-PSK modulation*

7.14.1. *Differential coding and decoding of the 8-PSK modulation*

The differential coding and decoding of the 8-PSK modulation is carried out according to the following general relationship:

$$\psi_n = \psi_{n-1} + \Delta\psi_n \qquad [7.123]$$

where the phase jump $\Delta\psi_n$ is defined for each three-bit sequence $v\ u\ c$ by Table 7.3.

$v\ u\ c$	000	001	011	010	110	111	101	100
$\Delta\psi_n$	0°	45°	90°	135°	180°	225°	270°	315°

Table 7.3. *Phase jump value as a function of the value of the three-bit sequence $v\ u\ c$*

We can then build the complete encoding and decoding Table 7.4, consisting of 64 states.

$Q_{n-1}I_{n-1}$ K_{n-1}	$v\ u\ c$	$Q_nI_nK_n$	$Q_{n-1}I_{n-1}$ K_{n-1}	$v\ u\ c$	$Q_nI_nK_n$
0 0 0	0 0 0	0 0 0	1 1 0	0 0 0	1 1 0
0 0 0	0 0 1	0 0 1	1 1 0	0 0 1	1 1 1
0 0 0	0 1 1	0 1 1	1 1 0	0 1 1	1 0 1
0 0 0	0 1 0	0 1 0	1 1 0	0 1 0	1 0 0
0 0 0	1 1 0	1 1 0	1 1 0	1 1 0	0 0 0
0 0 0	1 1 1	1 1 1	1 1 0	1 1 1	0 0 1
0 0 0	1 0 1	1 0 1	1 1 0	1 0 1	0 1 1
0 0 0	1 0 0	1 0 0	1 1 0	1 0 0	0 1 0
0 0 1	0 0 0	0 0 1	1 1 1	0 0 0	1 1 1
0 0 1	0 0 1	0 1 1	1 1 1	0 0 1	1 0 1
0 0 1	0 1 1	0 1 0	1 1 1	0 1 1	1 0 0
0 0 1	0 1 0	1 1 0	1 1 1	0 1 0	0 0 0
0 0 1	1 1 0	1 1 1	1 1 1	1 1 0	0 0 1

$Q_{n-1}I_{n-1}$ K_{n-1}	$v\,u\,c$	$Q_nI_nK_n$	$Q_{n-1}I_{n-1}$ K_{n-1}	$v\,u\,c$	$Q_nI_nK_n$
0 0 1	1 1 1	1 0 1	1 1 1	1 1 1	0 1 1
0 0 1	1 0 1	1 0 0	1 1 1	1 0 1	0 1 0
0 0 1	1 0 0	0 0 0	1 1 1	1 0 0	1 1 0
0 1 1	0 0 0	0 1 1	1 0 1	0 0 0	1 0 1
0 1 1	0 0 1	0 1 0	1 0 1	0 0 1	1 0 0
0 1 1	0 1 1	1 1 0	1 0 1	0 1 1	0 0 0
0 1 1	0 1 0	1 1 1	1 0 1	0 1 0	0 0 1
0 1 1	1 1 0	1 0 1	1 0 1	1 1 0	0 1 1
0 1 1	1 1 1	1 0 0	1 0 1	1 1 1	0 1 0
0 1 1	1 0 1	0 0 0	1 0 1	1 0 1	1 1 0
0 1 1	1 0 0	0 0 1	1 0 1	1 0 0	1 1 1
0 1 0	0 0 0	0 1 0	1 0 0	0 0 0	1 0 0
0 1 0	0 0 1	1 1 0	1 0 0	0 0 1	0 0 0
0 1 0	0 1 1	1 1 1	1 0 0	0 1 1	0 0 1
0 1 0	0 1 0	1 0 1	1 0 0	0 1 0	0 1 1
0 1 0	1 1 0	1 0 0	1 0 0	1 1 0	0 1 0
0 1 0	1 1 1	0 0 0	1 0 0	1 1 1	1 1 0
0 1 0	1 0 1	0 0 1	1 0 0	1 0 1	1 1 1
0 1 0	1 0 0	0 1 1	1 0 0	1 0 0	1 0 1

Table 7.4. *Complete differential 8-PSK coding and decoding table (64 states)*

7.14.2. Realization of the differential encoder and decoder: by Simulink simulation (MATLAB) and hardware implementation based on a ROM or EPROM memory

7.14.2.1. *Realization under Simulink*

The differential encoder equations Q_n, I_n, K_n are directly implemented with Simulink, as shown in Figure 7.42.

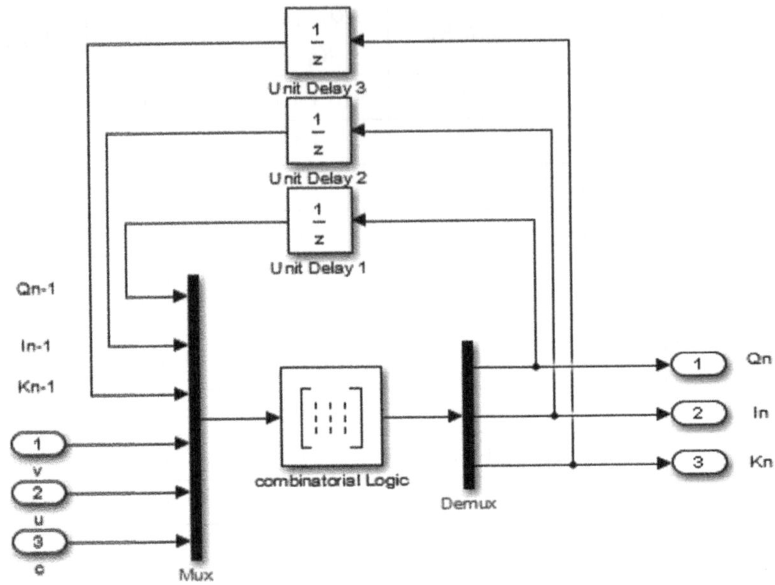

Figure 7.42. *Simulation of the differential 8-PSK encoder with Simulink*

Similarly, the equations v, u, c of the differential 8-PSK decoder are directly implemented with Simulink, as shown in Figure 7.43.

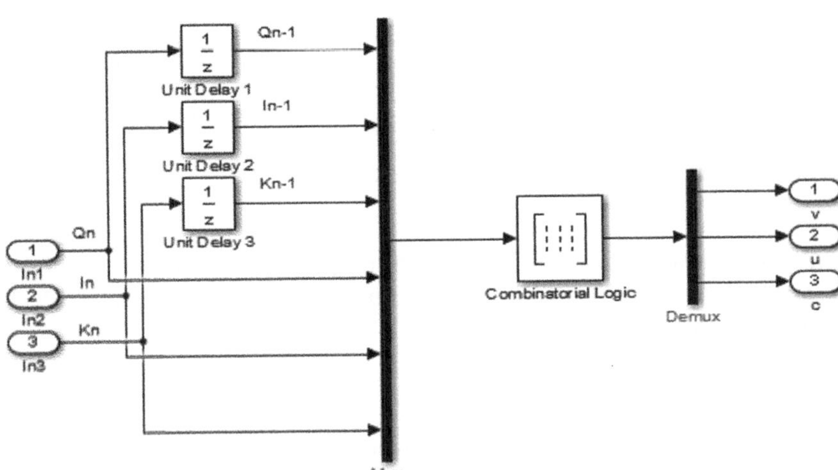

Figure 7.43. *Simulation of the differential 8-PSK decoder with Simulink*

7.14.2.2. *Implementation of the differential encoder and decoder by a ROM or EPROM memory*

One of the easiest ways to physically implement a look-up table is to use a ROM or EPROM, as shown in Figure 7.44. The achievement is based solely on the use of a circuit that implements a two-dimensional array of bits.

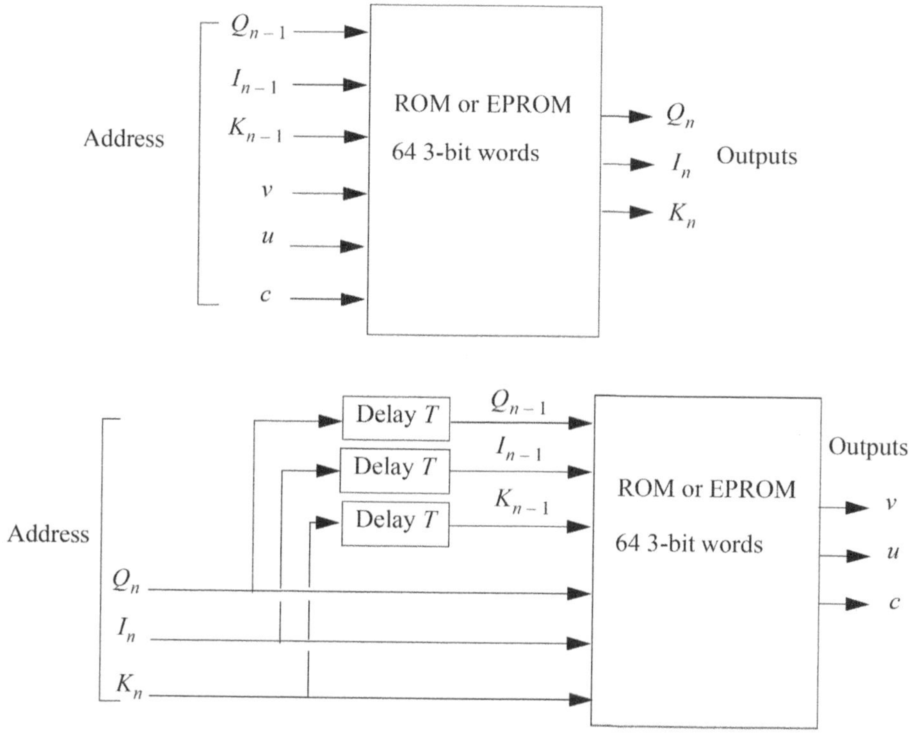

Figure 7.44. *Implementation of the differential encoder (top) and the decoder (bottom) by a ROM or EPROM memory*

7.14.2.3. *8-PSK transmitter and receiver including differential encoder and decoder*

In Figure 7.45, we give the diagram of the 8-PSK transmitter with the differential encoding, and in Figure 7.46, that of the 8-PSK receiver with the differential decoding.

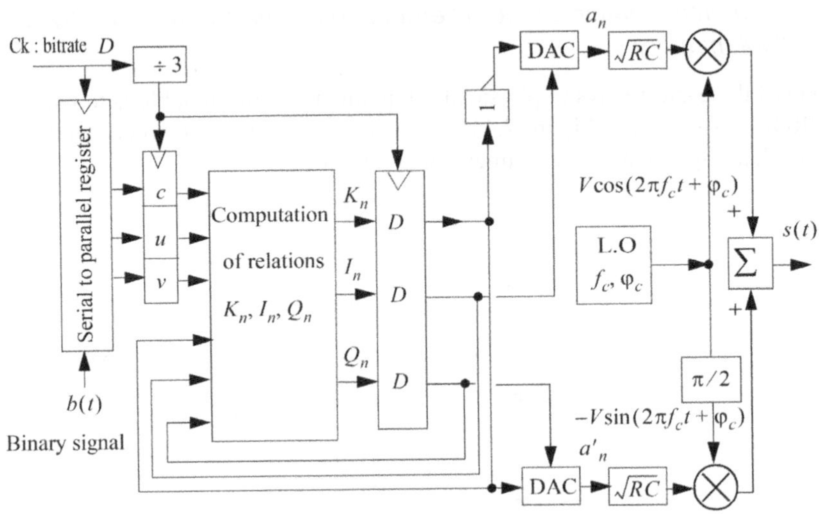

Figure 7.45. *Diagram of the 8-PSK transmitter including the differential encoder*

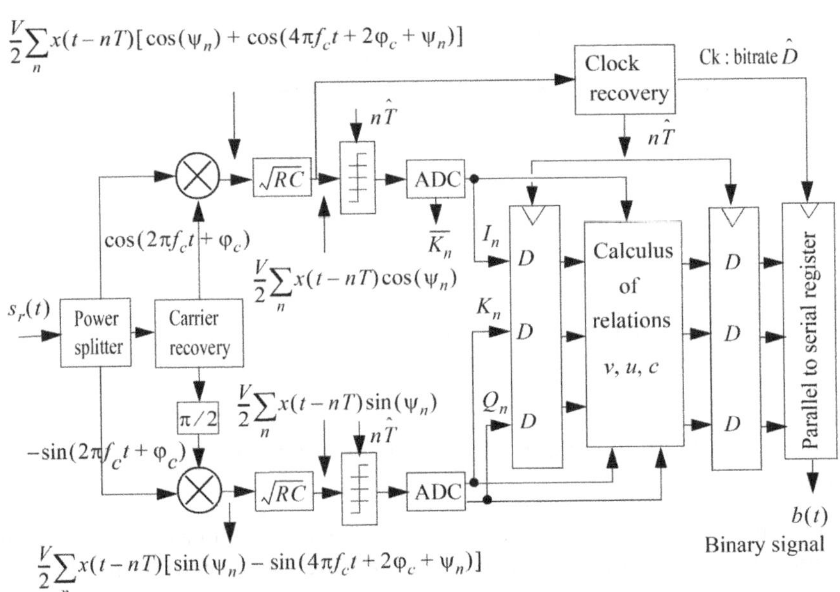

Figure 7.46. *Diagram of the 8-PSK receiver including the differential decoder*

7.14.2.4. *Spectral occupancy of the 8-PSK signal*

In the 8-PSK modulation, we have:

$$M = 2^3 = 8 \rightarrow D_S = D_Q = D_I = D_K = D/3 \qquad [7.124]$$

The frequency band occupied by the signal is maximum B_{Max}, when the bit sequence on channels Q, I, and K alternately changes between the 1 and 0 binary values (see Figure 7.47).

The highest fundamental frequency is equal to:

$$f_a = 1/T_a = 1/6T_b = D/6 \qquad [7.125]$$

and the Nyquist frequency is then:

$$f_N = 2f_a = D/3 = D_s \qquad [7.126]$$

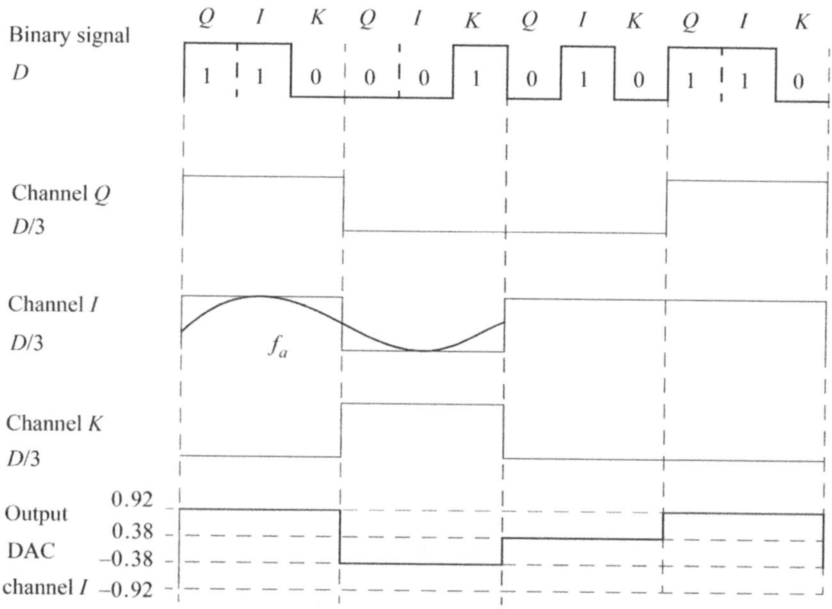

Figure 7.47. *Different channels for a binary sequence that changes alternately between binary values 1 and 0 and DAC output for channel I*

EXAMPLE.– Let us take a 8-PSK modulator, with $f_c = 70$ MHz and $D = 10\,\text{Mbit/s}$.

1) Calculate the symbol rate D_s, the fundamental frequency f_a, and the Nyquist frequency f_N.

2) Display the spectrum of the modulated signal $s(t)$.

3) compare the results obtained with modulators 16-QAM, QPSK and BPSK.

SOLUTION.–

1) The symbol rate is $D_s = D/3 = 3.333\,\text{Msymbol/s}$.

The fundamental frequency is $f_a = D/6 = 1.667$ MHz.

The Nyquist frequency is $f_N = 2f_a = 3.333$ MHz.

2) The signal $s(t)$ and its spectrum are, respectively, given by:

$$s(t) = \frac{V}{2}\{\cos[2\pi(f_c - f_a)t] + \cos[2\pi(f_c + f_a)t]\}$$

$$s(t) = \frac{V}{2}\{\cos[2\pi(68.333 \times 10^6)t] + \cos[2\pi(71.667 \times 10^6)t]\}$$

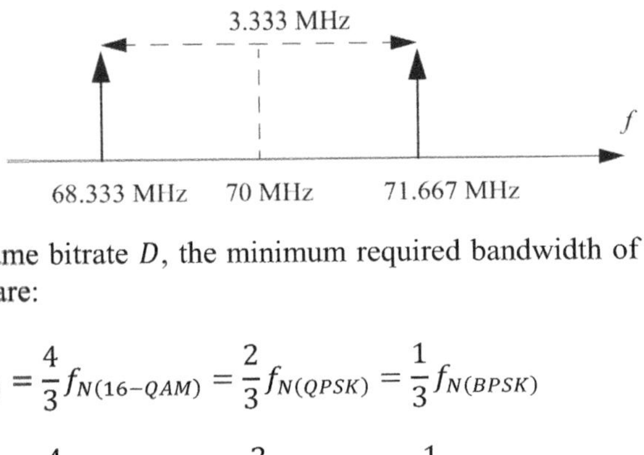

3) For the same bitrate D, the minimum required bandwidth of the channel and the symbol rate are:

$$f_{N(8-PSK)} = \frac{4}{3} f_{N(16-QAM)} = \frac{2}{3} f_{N(QPSK)} = \frac{1}{3} f_{N(BPSK)}$$

$$D_{s(8-PSK)} = \frac{4}{3} D_{s(16-QAM)} = \frac{2}{3} D_{s(QPSK)} = \frac{1}{3} D_{s(BPSK)}$$

NOTE.– For M > 8, it is preferable to use QAM modulation than a pure M-PSK modulation.

7.15. Performances of modulations in spectral occupancy and efficiency

In Table 7.5, we summarize the performances of the linear digital modulations, in terms of spectral occupancy and efficiency, given for a carrier frequency of 70 MHz.

Modulation	Theoritical spectral occupancy $f_N = 1/T = D_s$	Spectral efficiency $\eta(bit/s/Hz)$	Spectral occupancy (KHz/channel)
BPSK	*D* (bit/s)	1	68.27–73.56
QPSK	*D*/2	2	34.13–36.78
OQPSK	*D*	2	68.27–73.56
8-PSK	*D*/3	3	22.76–24.52
16-QAM	*D*/4	4	17.06–18.39
64-QAM	*D*/6	6	11.38–12.26
256-QAM	*D*/8	8	8.53–9.19

Table 7.5. *Performance of linear digital modulations in spectral occupancy and efficiency*

The spectral efficiency measures the number of bits transmitted per second and per Hertz. At constant bit error rate (BER) the bandwidth efficiency $\eta = f(E_b/b_0)$ increases as the energy E_b per bit increases.

Finally, note that the spectral occupancy differences indicated in the last column of Table 7.5 are related to the redundancy required for the transmission from the lowest hierarchical level (DT1: 2.048 Mbit/s or 30 telephone channels) to the highest (DT5: 564.992 Mbit/s or 7,680 telephone channels).

References

Assaf, R., El Assad, S., Harkous, Y. (2010). *Simulation et égalisation adaptative des canaux de transmission*. Éditions universitaires européennes, Nantes.

Barba, D. (2019). Cours de théorie et transmission de l'information. Polytech, Nantes.

Baudoin, G. (2002). *Radiocommunications numériques 1*. Dunod, Paris.

Bic, J.C., Duponteil, D., Imbeaux, J.C. (1986). *Éléments de communications numériques*. Dunod, Paris.

Caragata, D., El Assad, S., Luduena, M. (2015). An improved fragile watermarking algorithm for JPEG images. *International Journal of Electronics and Communications*, 69, 1783–1794.

Charbit, M. (2003). *Systèmes de communications et théorie de l'information*. Hermes-Lavoisier, Cachan.

Combes, P.F. (1996). *Micro-ondes : lignes, guides et cavités, cours et exercices*. Dunod, Paris.

Csillag, P. (1990). *Introduction aux codes correcteurs*. Ellipses, Paris.

De Cannière, C. and Preneel, B. (2006). Trivium specifications [Online]. Available at: http://www.ecrypt.eu.org/stream/p3ciphers/trivium/trivium_p3.pdf.

Dixon, R.C. (1994). *Spread Spectrum Systems with Commercial Applications*. John Wiley & Sons, Hoboken.

El Assad, S. (2001). Conception et simulation d'un modem MDP-8 cohérent et différentiel. *International Conference on Image and Signal Processing, ICISP 2001*, Agadir, 845–852.

El Assad, S. (2019). Cours de communications numériques : bases et techniques. Polytech, Nantes.

El Assad, S. and Barba, D. (2020). *Communications numériques 2 : travaux dirigés et pratiques*. ISTE Éditions, London.

El Assad, S., *et al.* (2003–2005). COBBALT : coopération ouest BAL en télécommunications. Internal report, Université virtuelle en Pays de la Loire.

El Assad, S., *et al.* (2006–2008). MASTER-FOAD, projet UVPL. Internal report, Université virtuelle en Pays de la Loire.

Fontolliet, P.G. (1983). *Systèmes de télécommunications*. Dunod, Paris.

Glavieux, A. and Joindot, M. (1996). *Communications numériques*. Masson, Paris.

Guglielmi, M. (2004). *Signaux aléatoires : modélisation, estimation, détection*. Hermes-Lavoisier, Paris.

Haykin, S. (1988). *Digital Communications*. John Wiley & Sons, Hoboken.

MacWilliams, F.J. and Stoane, N.J.A. (1977). *The Theory of Error Correcting Codes*. Bell Laboratories, Murray Hill.

MathWorks (2020). comm.KasamiSequence [Online]. Available at: https://fr.mathworks.com/help/comm/ref/comm.kasamisequence-system-object.html.

Maximov, A. and Biryukov, A. (2007). Two trivial attacks on Trivium. In *Selected Areas in Cryptography*, Adams, C., Miri, A. and Wiener, M. (eds). Springer-Verlag, Berlin.

Paar, C. and Oelzl, J. (2010). *Understanding Cryptography*. Springer-Verlag, Berlin.

Pätzold, M. (2002). *Mobile Fading Channels*. John Wiley & Sons, Hoboken.

Penaud, S. (2001). Etude des potentialités du chaos pour les systèmes de télécommunications : évaluation des performances de systèmes à accès multiples à répartition par les codes (CDMA) utilisant des séquences d'étalement chaotiques. Thesis, University of Limoges.

Proakis, J.G. (1995). *Digital Communications*. McGraw-Hill, New York.

Roubine, E. (1970). *Introduction à la théorie de la communication III*. Masson, Paris.

Sanchez, J. and Thioune, M. (2008). *UMTS*, 3rd edition. Hermes-Lavoisier, Cachan.

Servin, C. (2009). *Réseaux & télécoms*. Dunod, Paris.

Sklar, B. (1988). *Digital Communications*. Prentice Hall, Upper Saddle River.

Spataru, A. (1987). *Fondements de la théorie de la transmission de l'information*. Presses polytechniques et universitaires romandes, Lausanne.

Stein, M. (1991). *Les modems pour transmission des données*. Masson, Paris.

Tomasi, W. (1998). *Electronic Communications Systems*. Prentice Hall, Upper Saddle River.

Webb, W. and Hanzo, L. (1994). *Modern Quadrature Amplitude Modulation*. Pentech Press, London.

Index

S

T

U, W

Other titles from

iSTE

in

Information Systems, Web and Pervasive Computing

2020

CLIQUET Gérard, with the collaboration of BARAY Jérôme
Location-Based Marketing: Geomarketing and Geolocation

DE FRÉMINVILLE Marie
Cybersecurity and Decision Makers: Data Security and Digital Trust

GEORGE Éric
Digitalization of Society and Socio-political Issues 2: Digital, Information and Research

SEDKAOUI Soraya, KHELFAOUI Mounia
Sharing Economy and Big Data Analytics

2019

ALBAN Daniel, EYNAUD Philippe, MALAURENT Julien, RICHET Jean-Loup, VITARI Claudio
Information Systems Management: Governance, Urbanization and Alignment

AUGEY Dominique, with the collaboration of ALCARAZ Marina
Digital Information Ecosystems: Smart Press

BATTON-HUBERT Mireille, DESJARDIN Eric, PINET François
Geographic Data Imperfection 1: From Theory to Applications

BRIQUET-DUHAZÉ Sophie, TURCOTTE Catherine
From Reading-Writing Research to Practice

BROCHARD Luigi, KAMATH Vinod, CORBALAN Julita, HOLLAND Scott, MITTELBACH Walter, OTT Michael
Energy-Efficient Computing and Data Centers

CHAMOUX Jean-Pierre
The Digital Era 2: Political Economy Revisited

COCHARD Gérard-Michel
Introduction to Stochastic Processes and Simulation

DUONG Véronique
SEO Management: Methods and Techniques to Achieve Success

GAUCHEREL Cédric, GOUYON Pierre-Henri, DESSALLES Jean-Louis
Information, The Hidden Side of Life

GEORGE Éric
Digitalization of Society and Socio-political Issues 1: Digital, Communication and Culture

GHLALA Riadh
Analytic SQL in SQL Server 2014/2016

JANIER Mathilde, SAINT-DIZIER Patrick
Argument Mining: Linguistic Foundations

SOURIS Marc
Epidemiology and Geography: Principles, Methods and Tools of Spatial Analysis

TOUNSI Wiem
Cyber-Vigilance and Digital Trust: Cyber Security in the Era of Cloud Computing and IoT

2018

ARDUIN Pierre-Emmanuel
Insider Threats
(Advances in Information Systems Set – Volume 10)

CARMÈS Maryse
Digital Organizations Manufacturing: Scripts, Performativity and Semiopolitics
(Intellectual Technologies Set – Volume 5)

CARRÉ Dominique, VIDAL Geneviève
Hyperconnectivity: Economical, Social and Environmental Challenges
(Computing and Connected Society Set – Volume 3)

CHAMOUX Jean-Pierre
The Digital Era 1: Big Data Stakes

DOUAY Nicolas
Urban Planning in the Digital Age
(Intellectual Technologies Set – Volume 6)

FABRE Renaud, BENSOUSSAN Alain
The Digital Factory for Knowledge: Production and Validation of Scientific Results

GAUDIN Thierry, LACROIX Dominique, MAUREL Marie-Christine, POMEROL Jean-Charles
Life Sciences, Information Sciences

GAYARD Laurent
Darknet: Geopolitics and Uses
(Computing and Connected Society Set – Volume 2)

IAFRATE Fernando
Artificial Intelligence and Big Data: The Birth of a New Intelligence
(Advances in Information Systems Set – Volume 8)

LE DEUFF Olivier
Digital Humanities: History and Development
(Intellectual Technologies Set – Volume 4)

MANDRAN Nadine
Traceable Human Experiment Design Research: Theoretical Model and Practical Guide
(Advances in Information Systems Set – Volume 9)

PIVERT Olivier
NoSQL Data Models: Trends and Challenges

ROCHET Claude
Smart Cities: Reality or Fiction

SAUVAGNARGUES Sophie
Decision-making in Crisis Situations: Research and Innovation for Optimal Training

SEDKAOUI Soraya
Data Analytics and Big Data

SZONIECKY Samuel
Ecosystems Knowledge: Modeling and Analysis Method for Information and Communication
(Digital Tools and Uses Set – Volume 6)

2017

BOUHAÏ Nasreddine, SALEH Imad
Internet of Things: Evolutions and Innovations
(Digital Tools and Uses Set – Volume 4)

DUONG Véronique
Baidu SEO: Challenges and Intricacies of Marketing in China

LESAS Anne-Marie, MIRANDA Serge
The Art and Science of NFC Programming
(Intellectual Technologies Set – Volume 3)

LIEM André
Prospective Ergonomics
(Human-Machine Interaction Set – Volume 4)

MARSAULT Xavier
Eco-generative Design for Early Stages of Architecture
(Architecture and Computer Science Set – Volume 1)

REYES-GARCIA Everardo
The Image-Interface: Graphical Supports for Visual Information
(Digital Tools and Uses Set – Volume 3)

REYES-GARCIA Everardo, BOUHAÏ Nasreddine
Designing Interactive Hypermedia Systems
(Digital Tools and Uses Set – Volume 2)

SAÏD Karim, BAHRI KORBI Fadia
Asymmetric Alliances and Information Systems:Issues and Prospects
(Advances in Information Systems Set – Volume 7)

SZONIECKY Samuel, BOUHAÏ Nasreddine
Collective Intelligence and Digital Archives: Towards Knowledge Ecosystems
(Digital Tools and Uses Set – Volume 1)

2016

BEN CHOUIKHA Mona
Organizational Design for Knowledge Management

BERTOLO David
Interactions on Digital Tablets in the Context of 3D Geometry Learning
(Human-Machine Interaction Set – Volume 2)

BOUVARD Patricia, SUZANNE Hervé
Collective Intelligence Development in Business

EL FALLAH SEGHROUCHNI Amal, ISHIKAWA Fuyuki, HÉRAULT Laurent, TOKUDA Hideyuki
Enablers for Smart Cities

FABRE Renaud, in collaboration with MESSERSCHMIDT-MARIET Quentin, HOLVOET Margot
New Challenges for Knowledge

GAUDIELLO Ilaria, ZIBETTI Elisabetta
Learning Robotics, with Robotics, by Robotics
(Human-Machine Interaction Set – Volume 3)

HENROTIN Joseph
The Art of War in the Network Age
(Intellectual Technologies Set – Volume 1)

KITAJIMA Munéo
Memory and Action Selection in Human–Machine Interaction
(Human–Machine Interaction Set – Volume 1)

LAGRAÑA Fernando
E-mail and Behavioral Changes: Uses and Misuses of Electronic Communications

LEIGNEL Jean-Louis, UNGARO Thierry, STAAR Adrien
Digital Transformation
(Advances in Information Systems Set – Volume 6)

NOYER Jean-Max
Transformation of Collective Intelligences
(Intellectual Technologies Set – Volume 2)

VENTRE Daniel
Information Warfare – 2nd edition

VITALIS André
The Uncertain Digital Revolution
(Computing and Connected Society Set – Volume 1)

2015

ARDUIN Pierre-Emmanuel, GRUNDSTEIN Michel, ROSENTHAL-SABROUX Camille
Information and Knowledge System
(Advances in Information Systems Set – Volume 2)

BÉRANGER Jérôme
Medical Information Systems Ethics

BRONNER Gérald
Belief and Misbelief Asymmetry on the Internet

IAFRATE Fernando
From Big Data to Smart Data
(Advances in Information Systems Set – Volume 1)

KRICHEN Saoussen, BEN JOUIDA Sihem
Supply Chain Management and its Applications in Computer Science

NEGRE Elsa
Information and Recommender Systems
(Advances in Information Systems Set – Volume 4)

POMEROL Jean-Charles, EPELBOIN Yves, THOURY Claire
MOOCs

SALLES Maryse
Decision-Making and the Information System
(Advances in Information Systems Set – Volume 3)

SAMARA Tarek
ERP and Information Systems: Integration or Disintegration
(Advances in Information Systems Set – Volume 5)

2014

DINET Jérôme
Information Retrieval in Digital Environments

HÉNO Raphaële, CHANDELIER Laure
3D Modeling of Buildings: Outstanding Sites

KEMBELLEC Gérald, CHARTRON Ghislaine, SALEH Imad
Recommender Systems

MATHIAN Hélène, SANDERS Lena
Spatio-temporal Approaches: Geographic Objects and Change Process

PLANTIN Jean-Christophe
Participatory Mapping

VENTRE Daniel
Chinese Cybersecurity and Defense

2013

BERNIK Igor
Cybercrime and Cyberwarfare

CAPET Philippe, DELAVALLADE Thomas
Information Evaluation

LEBRATY Jean-Fabrice, LOBRE-LEBRATY Katia
Crowdsourcing: One Step Beyond

SALLABERRY Christian
Geographical Information Retrieval in Textual Corpora

2012

BUCHER Bénédicte, LE BER Florence
Innovative Software Development in GIS

GAUSSIER Eric, YVON François
Textual Information Access

STOCKINGER Peter
Audiovisual Archives: Digital Text and Discourse Analysis

VENTRE Daniel
Cyber Conflict

2011

BANOS Arnaud, THÉVENIN Thomas
Geographical Information and Urban Transport Systems

DAUPHINÉ André
Fractal Geography

LEMBERGER Pirmin, MOREL Mederic
Managing Complexity of Information Systems

STOCKINGER Peter
Introduction to Audiovisual Archives

STOCKINGER Peter
Digital Audiovisual Archives

VENTRE Daniel
Cyberwar and Information Warfare

2010

BONNET Pierre
Enterprise Data Governance

BRUNET Roger
Sustainable Geography

CARREGA Pierre
Geographical Information and Climatology

CAUVIN Colette, ESCOBAR Francisco, SERRADJ Aziz
Thematic Cartography – 3-volume series
Thematic Cartography and Transformations – Volume 1
Cartography and the Impact of the Quantitative Revolution – Volume 2
New Approaches in Thematic Cartography – Volume 3

LANGLOIS Patrice
Simulation of Complex Systems in GIS

MATHIS Philippe
Graphs and Networks – 2nd edition

THERIAULT Marius, DES ROSIERS François
Modeling Urban Dynamics

2009

BONNET Pierre, DETAVERNIER Jean-Michel, VAUQUIER Dominique
Sustainable IT Architecture: the Progressive Way of Overhauling Information Systems with SOA

PAPY Fabrice
Information Science

RIVARD François, ABOU HARB Georges, MERET Philippe
The Transverse Information System

ROCHE Stéphane, CARON Claude
Organizational Facets of GIS

2008

BRUGNOT Gérard
Spatial Management of Risks

FINKE Gerd
Operations Research and Networks

GUERMOND Yves
Modeling Process in Geography

KANEVSKI Michael
Advanced Mapping of Environmental Data

MANOUVRIER Bernard, LAURENT Ménard
Application Integration: EAI, B2B, BPM and SOA

PAPY Fabrice
Digital Libraries

2007

DOBESCH Hartwig, DUMOLARD Pierre, DYRAS Izabela
Spatial Interpolation for Climate Data

SANDERS Lena
Models in Spatial Analysis

2006

CLIQUET Gérard
Geomarketing

CORNIOU Jean-Pierre
Looking Back and Going Forward in IT

DEVILLERS Rodolphe, JEANSOULIN Robert
Fundamentals of Spatial Data Quality

Printed and bound by CPI Group (UK) Ltd, Croydon, CR0 4YY

16/09/2026

14963077-0006